U0121125

湿地光影
丛书
COLLECTION OF
WETLANDS IN IMAGES

微距生灵

CLOSE-UP VIEWS OF CREATURES IN WETLANDS

袁明辉
Yuan Minghui
著/摄

中国林業出版社
China Forestry Publishing House

图书在版编目（CIP）数据

微距生灵 / 袁明辉著、摄. -- 北京 : 中国林业出版社，2022.10
（湿地光影丛书）
ISBN 978-7-5219-1888-5

Ⅰ. ①微… Ⅱ. ①袁… Ⅲ. ①沼泽化地－普及读物
Ⅳ. ①P931.7-49

中国版本图书馆CIP数据核字(2022)第181871号

出 版 人：成　吉
总 策 划：成　吉　王佳会
策　　划：杨长峰　肖　静
责任编辑：袁丽莉　肖　静
宣传营销：张　东　王思明
英文翻译：袁明辉
特约编辑：田　红
图片编辑：崔　林
装帧设计：崔　林（依丹设计）

出版发行：中国林业出版社（100009#北京市西城区刘海胡同7号）
http://www.forestry.gov.cn/lycb.html
电话：（010）83143577
E-mail：forestryxj@126.com
印刷：北京雅昌艺术印刷有限公司
版次：2022年10月第1版
印次：2022年10月第1次
开本：787mm × 1092mm　1/12
印张：18
字数：90千字
定价：320.00元

序言 Foreword

当我第一次看到袁明辉的照片，就立刻看出他是位技艺高超的摄影师。由于我是一名专业的摄影师、书籍和杂志的设计师，并且我经常在国际摄影比赛中担任评委，所以每年都有成千上万的自然摄影作品在我眼前掠过，但他的照片与众不同，完美地融合了记录和诠释自然的功能，并具有无可挑剔的构图。他的风格结合了伟大的东方审美情感和欧洲自然摄影的最佳传统。

我为意大利自然摄影杂志《Asferico》采访他时，我意识到这些照片与摄影师的内心哲学完美契合，是他面对大自然真实情感的反映。如果仔细思考，这些图片几乎是极其简单的，并且在大多数情况下，它们是我们每个人在大自然中漫步时都可以轻松观察到的常见主题。但作者超乎寻常的敏锐度却将这些主题变成了真正的艺术作品。虽然罕见主题的画面要引起观察者的惊叹和激动并不容易，但要在视觉上成功地呈现所有人眼皮底下的常见主题更是难上加难。我坚信，在自然界中，水生植物简单的一片叶子与隐藏在难以到达的森林中的稀有主题具有相同的重要性。

在我们的谈话中，袁明辉谈到了自然主体的“尊严”，我认为这个特定的词非常重要。我相信今天自然摄影师的职责之一，应该是为那些看似无关紧要的小生命赋予尊严。

今天的自然摄影比过去更容易了。通过相机和摄影拍摄技术可以获得完美的高分辨率图像，一些镜头和配件的价格也更加合理。即使在我们星球的一些偏远之地，摄影师也可以轻松携带相机旅行和拍照。观察其他摄影师的作品并在某些方面“复制”他们的风格也很容易。但商店永远不卖给你的是你的内心看待自然的方式。对自然主题的敏锐度和美感不能以任何方式获取。你要么拥有它们，要么没有。摄影不是相机、镜头、光圈、速度和三脚架的简单问题，而是一个内在感性和哲学的问题，也是一个摄影师和另一个摄影师之间的区别。

直到今天，在这本精美的书中，读者有机会欣赏作者的一些令人惊叹的摄影作品。我祝愿我的朋友袁先生今后一切顺利。

专业自然摄影师和作家
国际自然摄影赛事评委
［意］扬尼斯•施内佐斯

When I watched Yuan Minghui's pictures for the first time, it was immediately clear that he was a photographer of great skill. Since I am a professional photographer, designer of books / magazines and I'm often a judge in photographic competitions, thousands of nature pictures pass under my eyes every year, but his photographs are different, a perfect fusion of documentation, interpretation and impeccable composition. His style combines a great oriental aesthetic sensibility with the best tradition of European nature photography.

When, a few months later, I interviewed him for the Italian nature photography magazine *Asferico*, I realized that those pictures coincided perfectly with the photographer's inner philosophy and they were the real and direct result of his emotions in front of the creations of nature.

If you reflect carefully, these pictures are of an almost extreme simplicity and in most cases they are rather common subjects that each of us can easily observe while wandering in nature; but the great sensibility of the author ensures that subjects become real artistic creations. While it is not easy to surprise and thrill an observer with an image of a rare subject, it is doubly difficult to visually render successfully a common subject that almost everyone has under their eyes. I firmly believe that in nature a simple leaf of an aquatic plant has the same importance of a rare subject hidden in hard to reach forests.

During our conversation, the author talked about "dignity" of natural subjects, and I think this specific word is of great importance. I believe that one of the duties of the nature photographer today should be to give dignity to small and apparently irrelevant subjects.

Today nature photography has become easier than in the past. The cameras produce high resolution images technically perfect, lenses and accessories are reasonably priced, photographers can move and take pictures easily even in remote places of our planet. It is also extremely simple to observe the work of other photographers and "copy" their style in some way. But what shops will never can sell you is the way your inner world looks at nature. Sensitivity and aesthetic sense towards natural subjects cannot be acquired in any way. You either have them or you don't. Photography it is not a simple matter of cameras, lenses, apertures, speeds and tripods: it is a question of inner sensibility and philosophy, this is the difference between one photographer and another.

In this beautiful book, readers have the opportunity to observe some of the author's stunning photographic work until today. I wish to my friend Yuan every success in the future.

Professional nature photographer and writer

Ioannis Schinezos

前言
Preface

清晨，我被手机的音乐铃声中断了梦境。想到今天要去湿地拍摄黎明的露珠和清晨的昆虫，立即抖擞精神，准备开始一天的工作。在我二十多年的自然观察和拍摄中，我始终对自然生态摄影师的职业充满了激情，更准确点说，是美丽的武汉东湖湿地给了我创作灵感，让我乐此不疲。

我依然记得童年时，妈妈带我在武汉郊外的池塘边采莲蓬，爸爸带我在南湖边看落霞与鸥鹭齐飞，姐姐和我在夕阳下追逐蜻蜓的情景。长大后，我想把童年的美好和那些野生动物朋友一一记录，不想忘记它们，因为最美好的时光总是流逝得最快。武汉的湿地里生活着很多的野生生命，它们会受到人和环境变化的侵扰，但是仍然在繁衍生息。不管四季怎样变迁，不管疫情如何肆虐，它们依然坚强地活着，因为这里是生它们、养它们的地方。

对于我而言，自然生态摄影已经成为一种使命——拍摄到更具深意的影像，向人们展示这些平凡生命的命运。这些平凡生命的美丽与尊严，就像我们老百姓的质朴与真实。我们应该带着人文关怀的理解去与自然和解，去展示这个星球上最容易被人遗忘的角落，去展示关注环境保护、关爱湿地的成果。

当接到中国林业出版社的邀约，为《关于特别是水禽栖息地的国际重要湿地公约》（简称《湿地公约》）第十四届缔约方大会出一本画册，我感到非常荣幸。缔约方大会于2022年11月在中国湖北省武汉市举办，这也是中国首次承办该国际会议。武汉是我的家乡，武汉承办缔约方大会，有助于进一步展示我国促进经济、社会与环境协调发展的负责任大国形象，是强化“一带一路”国家生态交流与合作的重要契机。

保护湿地就是保护我们的家园，保护我们自身，自然生态摄影最终能惠及湿地的动植物本身更具有意义。湿地里孕育着无限生机，蛙声虫鸣，兽踪鸟影……湿地的狂想曲开始了前奏。我想象着自然生态摄影与音乐的结合，湿地里生灵命运的一幕幕浮现在眼前，仿佛在看电影回放……沸腾的生活已经开始！

In the morning, I was interrupted by the music ring of my mobile phone. Thinking of going to the wetland today to shoot dewdrops at dawn and insects in the morning, I immediately refreshed myself and prepared to start the day’s work. In my more than 20 years of natural observation and shooting, I have always been full of passion for the career of natural photographer. To be more precise, it is the beautiful Wuhan East Lake wetland that gives me creative inspiration and keeps me happy.

I still remember when I was a child, my mother took me to pick lotus canopies by the pond in the suburbs of Wuhan; my father took me to watch the sunset and gulls take off by the South Lake; My sister and I chased dragonflies in the sunset. When I grow up, I want to record the beauty of my childhood and those wild animal friends one by one, and I don’t want to forget them, because the best time always passes the fastest. There are many wild life in the wetlands of Wuhan. They will be disturbed by human and environmental changes, but they are still multiplying. No matter how the seasons change, no matter how rampant the epidemic is, they are still strong to live, because this is where they are born and raised.

For me, natural photography has become a mission, that is to capture more meaningful images to show people the fate of these ordinary creatures. The beauty and dignity of these ordinary creatures is just like the simplicity and life of our people. With the understanding of humanistic care, we should reconcile with nature, show the most easily forgotten corner of the planet, and show the results of paying attention to environmental protection and caring for wetlands.

When I received the invitation from China Forestry Publishing House to publish album about the wetland for the 14th Conference of the Parties to the *Convention on Wetlands of International Importance Especially as Warterfowl Habitat* (*Convention on Wetlands* for short), I felt very honored. Coincidently, the conference of the parties will be held in Wuhan, Hubei Province, China in November 2022, which is the first time that China will host the international conference. Wuhan is my hometown. The hosting of the conference of the parties in Wuhan will help further display the image of China as a responsible big country promoting the coordinated development of economy, society and environment, and is an important opportunity to strengthen ecological communications and cooperation among the "the Belt and Road" countries.

Protecting wetlands is to protect our homes and ourselves. Natural photography can ultimately benefit the animals and plants in wetlands, which is more meaningful. There is infinite vitality in the wetland, the sound of frogs and insects, the trace of animals and the shadow of birds··· The Rhapsody of the wetland has begun its prelude. I imagine the combination of natural photography and music. The scenes of the fate of creatures in the wetland emerge in front of me, which seems to be watching movie playback··· The boiling life has begun!

目录
Contents

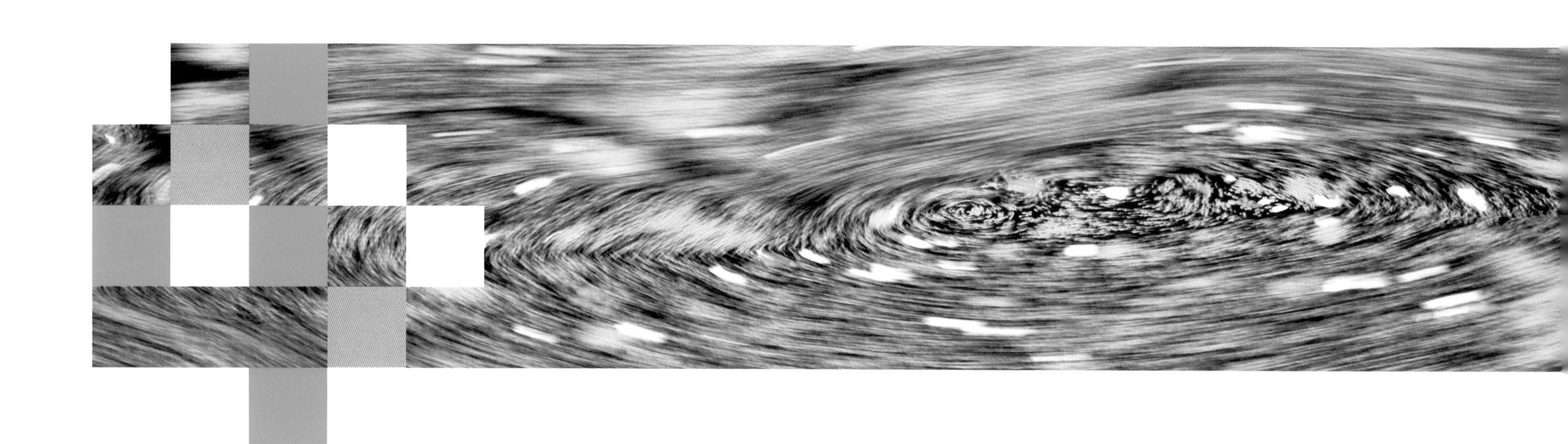

上善若水

Great Good Is Said to Be Like Water

水哺育了世间万物，却不向万物索取。水向上化为云雾，向下化作雨露。水有着最为自由的本身，聚可云结雨，散可无影无踪，飘忽于天地之间。水是世界上最温柔的，也是大自然里最顽强的。

Water nurtures all things in the world, but does not ask for anything from all things. Water turns upward into clouds and downward into rain and dew. Water has the most free itself. Gather clouds and rain, scatter without a trace, and float in the heaven and earth. Water is the gentlest in the world and the most tenacious in nature.

武汉东湖湿地的初秋，一场暴雨之后，天空开始变亮。

我在武汉市郊外的树林中发现了几株野葡萄藤缠绕在树干上。

这些盘绕弯曲的藤蔓就像五线谱上的高音谱号。

随着藤蔓上的雨滴滴落，我聆听到了大自然中美丽的声音。

此刻，来自音乐的灵感和藤蔓之间形成了一个意象的连接。

我注意到黄昏时斜射的光线从树林背后照过来。

我选择用大光圈虚化背景，同时保持镜头焦平面的平衡，让这个高音谱号的形式从杂乱的背景中独立出来。

我拍摄了大概一百多张照片，最后获得了我想要的影像。

用音乐的形式来表现阳光、水和空气这三个生命最抽象的元素，是音乐的印象和自然元素的完美结合。

In the early autumn of Wuhan East Lake wetland, the sky began to brighten after a rainstorm.
I found several wild vines winding around the trunk in the woods near the wetland.
These winding vines are like the treble clef on the staff. As the rain dripped on the vines, I heard the beautiful sound of nature.
At that moment, an image connection was formed between the inspiration from music and the vine.
I noticed the oblique light coming from behind the trees at dusk.
I chose to use a large aperture to virtualize the background, while maintaining the balance of the lens focal plane,
so that the form of treble clef could be independent from the chaotic background.
I took about a hundred photos, and finally I got the image I wanted.
The three most abstract elements of life, sunshine, water and air, are expressed in the form of music,
which is the perfect harmony between the impression of music and natural elements.

自然的和声
Natural Harmony

武汉东湖湿地的一处水潭上漂满了浮萍，
树上的花瓣飘落在水潭中。
花瓣和浮萍在水流的作用下形成一个漩涡。
我能想象在绿色背景下不同色彩的线条带来的抽象，
多彩的线条象征不同生命时光的流逝。

In spring, the surface of a pool in Wuhan East Lake wetland is full of duckweed.
This season is the flower season in Wuhan.
The petals fell into the pool.
The petals and duckweed form a vortex under the action of water flow.
I can imagine abstractions brought by different colored lines under the green background.
Colorful lines symbolize the passage of different life times.

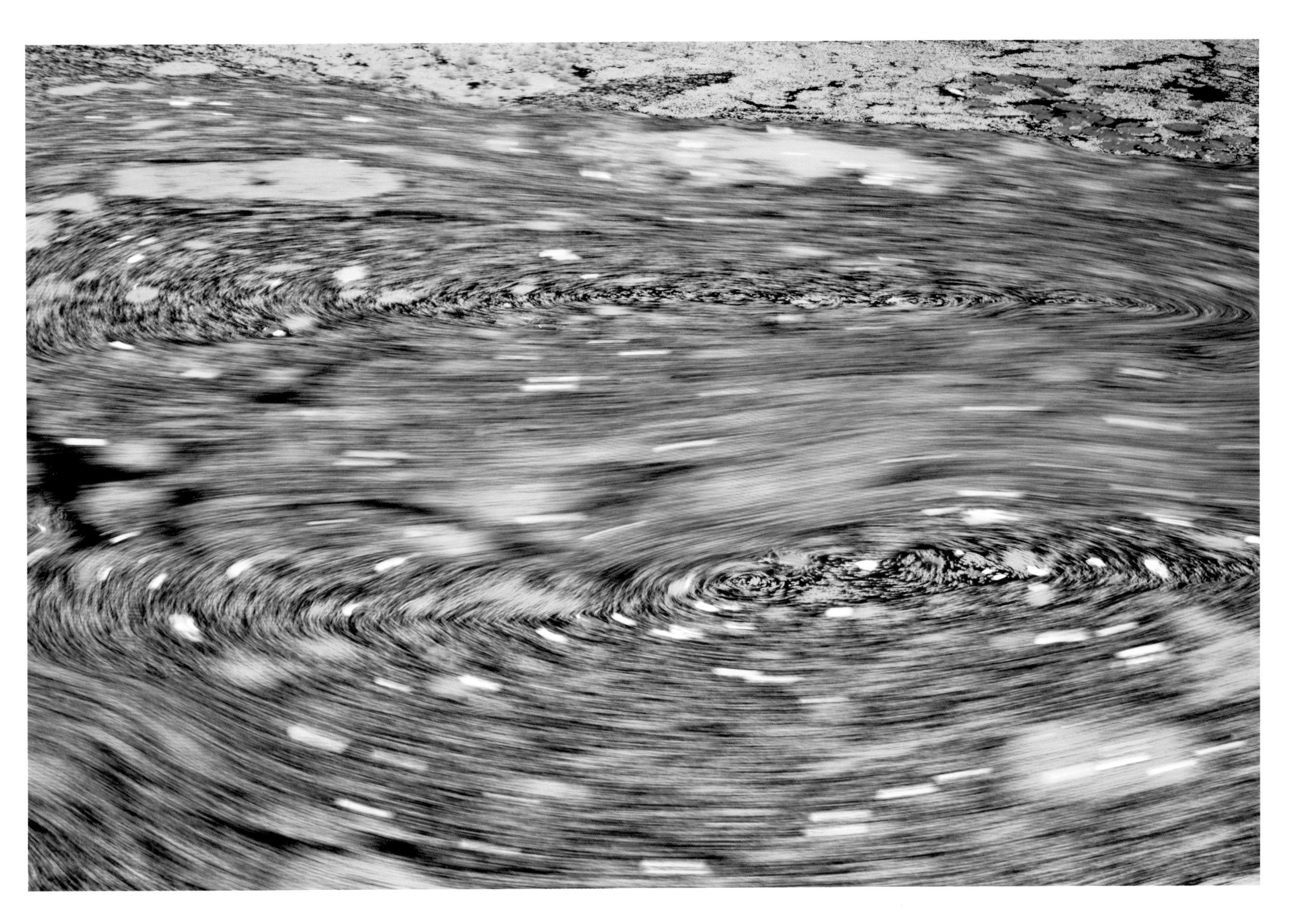

花瓣旋涡
Petal Vortex

槐叶苹是一种浮水性蕨类植物。

它生于水田、池塘的温暖、无污染的静水水域上。

在武汉东湖湿地一个夏季雨后的中午，

槐叶苹上的水滴在阳光下像钻石一般闪闪发光。

我想象槐叶苹像贝壳一样夹着一颗硕大的珍珠。

Salvinia natans is a floating fern aquatic plant.

It lives in the warm, unpolluted still waters of paddy fields and ponds.

In the summer of Wuhan East Lake wetland,

The water droplets on the leaves of *Salvinia natans* glitter like diamonds in the sun.

I imagine *Salvinia natans* holding a large pearl like a shell.

散落的钻石
Scattered Diamonds

在武汉东湖湿地，我经常能看见生长的蒲公英的小黄花。

蒲公英的种子都带着一把“降落伞”，在有风的时候会离开故土，随风飞舞。

我发现四颗蒲公英的种子落在小池塘的污水上。

四颗蒲公英的种子在水面上连接在一起。

贴着水面看上去，水面的灰尘在镜头里形成了梦幻般的白色光点。

蒲公英种子和它们的倒影构成了一个整体，就像四个舞者穿着芭蕾舞裙在舞台上跳舞。

In Wuhan East Lake wetland,

I can often see small yellow flowers of dandelion (*Taraxacum mongolicum*) growing.

Each seed of dandelion carries a parachute. When the wind blows, the seeds will leave their homeland and fly with the wind.

I found four dandelion seeds on the sewage of a small pond.

Four dandelion seeds join together on the water.

Looking close to the water surface, the dust on the water surface forms a dreamy white spot in the lens.

The dandelion seeds and their reflections form a whole, like four dancers dancing ballet on stage in tutus.

芭蕾女孩
Ballerina Girls

在武汉东湖湿地的夏季，总会有很多发现。

当我发现一只小青蛙趴在荷叶边缘时，我轻轻地从背后接近它。

我使用长焦微距镜头，这样可以离青蛙远一点拍摄。

青蛙的姿态像是人的坐姿，它的两只手搭在荷叶边缘，

荷叶就像是一个餐厅里的圆桌，它在等待粗心的昆虫来喝荷叶中的水。

青蛙是聪明的，它知道利用荷叶上的水来捕捉昆虫。

In the summer of Wuhan East Lake wetland,

there will always be many discoveries.

When I found a little frog on the edge of the lotus leaf, I approached it gently from behind.

I used a telephoto macro lens so I could shoot away from the frog.

The frog's posture is like a person's sitting posture.

Its two "hands" are on the edge of the lotus leaf.

The lotus leaf is like a round table in a restaurant.

It is waiting for careless insects to drink the water in the lotus leaf.

The frog is smart. It knows how to use the water on the lotus leaf to catch insects.

等待进餐
Waiting for Dinner

雨后的武汉东湖湿地，
满江红漂浮在池塘的水面，野荷也在池塘中生长。
荷叶浮在满江红之上，
一滴雨水在未完全展开的荷叶中，犹如一只眼睛。
雨水是悲伤或喜悦的泪水。

In Wuhan East Lake wetland after rain,
the red duckweed floats on the surface of the pond,
and the wild lotus also grows in the pond.
Lotus leaves float on red duckweed,
and a drop of rain in the lotus leaf that is not fully unfolded, like an eye.
Rain is tears of sadness and joy.

眼中的一滴泪
A Tear in the Eye

武汉东湖湿地的春季是苔藓生长的旺季。
一对苔藓的孢子囊在雨后相互依靠，
就像经历风雨磨难的两个人相依为命。
它们身上挂着的雨珠仿佛久别重逢后幸福的泪水。

The spring of Wuhan East Lake wetland is the peak season for moss growth.
A pair of moss sporangia depend on each other after the rain.
They are like two people who have experienced the hardships of wind and rain.
The raindrops hanging on them are like tears of happiness after a long separation and reunion.

风雨后相拥
Embracing Each Other after Rain

一个夏天的下午，在武汉市黄陂区的一条小溪里，
我发现一只大水黾正在水面上吃着一只雄性透顶单脉色蟌。
湿地生态系统通常处于陆地生态系统和水生生态系统的过渡地带，
适者生存法则经常在这里上演。

One summer afternoon, in a stream in Huangpi District, Wuhan City,
I found a large uater strider (*Aquarlus elongatus*) eating a male single-veined pierogi (*Matrona basilaris*) on the water.
Wetland ecosystem is usually in the transition area between terrestrial ecosystem and the aquatic ecosystem,
where the law of the survival of the fittest is often staged.

水面的猎手

Hunter on the Water

武汉东湖湿地的夏季，

新生的睡莲叶刚钻出池塘的水面。

睡莲嫩叶卷曲的形状像是一颗红色的爱心。

我寻觅到两片心形睡莲叶靠得很近的造型。

卷曲的睡莲叶就像是两颗心灵彼此为爱相互陪伴。

In the summer of Wuhan East Lake wetland,

the newborn water lily leaves have just drilled out of the water of the pond.

The curly shape of the tender leaves of water lily looks like a red heart.

I found two heart-shaped water lily leaves close to each other.

Curly water lily leaves are like two hearts protecting and accompanying each other for love.

漂浮的心
Floating Heart

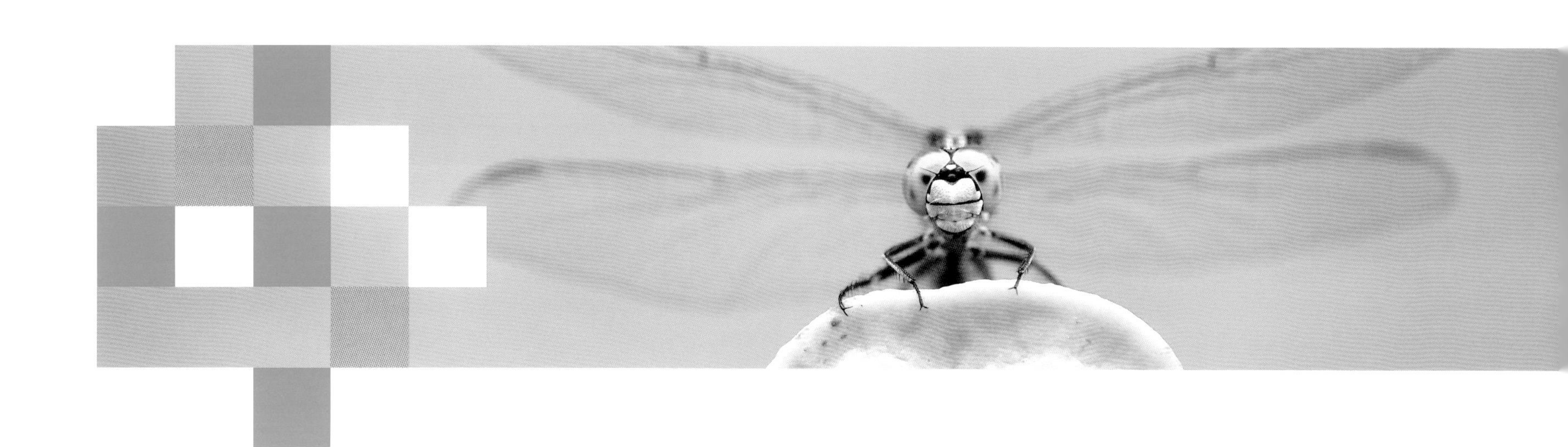

物竞天择

Natural Selection

生物互相竞争，能适应生存者被选择存留下来。我们生存于世是一个奇迹。自然是自然的，存在着天然的平衡，人也是这平衡中的一环。尊重并善待生命就是善待我们人类自己。

Creatures compete with each other, and those who can adapt to life are selected to survive. It is a miracle that we live in the world. Nature is natural, where there is a natural balance, and man is also a part of this balance. To respect and treat life well is to treat ourselves well.

在湖北仙岛湖附近，

一只枯叶蛾停留在树的断枝上，

完美地融入了它生活的自然环境，让它的天敌难以发现。

如果人类不能适应周围变化的环境，很有可能走向灭亡。

A lappet moth (Lasiocampidae) stays on a broken branch of the tree near Xiandao Lake, Hubei Province.

It is perfectly integrated into the natural environment where it lives,

making it difficult for its natural enemies to find.

If the human beings cannot adapt to the changing environment around them, they are likely to perish.

融入枯叶中
Integrate into Dead Leaves

在武汉东湖湿地的一个深秋时节，一些叶蜂的幼虫正在吃月季枝上的残叶。

一只全身嫩黄色的叶蜂幼虫在逆光下吸引了我的注意。

原来这只叶蜂幼虫刚蜕皮，它安静地待在叶片洞中。

天冷了，留给毛毛虫继续生长的时日已经不多。

在寒冬来临前，毛毛虫完成了一次生命的蜕变。

看到这条弱小的毛毛虫，我内心有一种感动：

生命在哪里结束，就会在哪里开始。

毛毛虫的理想也许就是快快长大，再次完成生命的蜕变。

In Wuhan East Lake wetland, one late autumn season,

some sawflies (Tenthredinidae) larvae were eating some residual leaves on rose branches.

A tender yellow sawfly larva attracted my attention in the backlight.

It turned out that the sawfly larva had just molted, and it stayed quietly in the leaf hole.

It's cold, and there's not much time left for caterpillars to continue to grow.

Before the cold winter comes, the caterpillar has completed a life transformation.

Seeing this weak caterpillar, I was moved: where life ends, it will start.

Caterpillar's ideal may be to grow up quickly and complete the transformation of life again.

毛毛虫的理想
Caterpillar’s Ideal

夏日的傍晚，

我发现一丛美人蕉叶片上有一排被虫子咬得大小不同，但形状相似的洞。

几乎同时，在叶片洞中的一只蹦蝗看到我之后躲在了洞旁边。

我长期在野外观察各种昆虫，对它们的习性比较了解。

我知道这只蹦蝗在感觉危险过去后又会出现在美人蕉叶片的洞中。

于是，我跪在这丛美人蕉边等待了大约一刻钟。

当蹦蝗再次出现在叶片洞时，我连续按下快门。

画面中的蹦蝗像是和我在玩捉迷藏的游戏，而我是游戏最后的赢家。

此刻的蹦蝗又像是为我打开了一扇窗户，

它在与我进行心灵的沟通。

On a summer evening,

I found a row of holes with different sizes but similar shapes bitten by insects on a bunch of Canna (*Canna indica*) leaves.

Almost at the same time, a locust (*Sinopodisma* sp.) in the leaf hole saw me and hid next to the hole.

I have been observing various insects in the field for a long time and have a better understanding of their habits,

so I knew this locust would appear in the hole of Canna leaves after feeling the danger had passed.

So I knelt by the Canna bush and waited for about a quarter of an hour.

When the locust appeared in the leaf hole again, I pressed the shutter continuously.

The locusts in the picture are playing hide and seek with me, and I am the last winner of the game.

At the moment, the locust seems to have opened a window for me. It is communicating with me.

捉迷藏
Hide and Seek

一只花蟹蛛在铁线莲种子上守候昆虫。
当它觉察到我的相机靠近之后，就弯曲它的前足向我示威，
就像一个人在展示他的肱二头肌，表示它很强大。
花蟹蛛身体太小了，为了拍摄得更大些，我继续靠近它拍摄。
花蟹蛛看到警告不起作用，于是翻身准备逃跑。
花蟹蛛展示出了它肚子上的像小丑面相的文身。

A crab spider (*Xysysticus* sp.) waits for insects on seeds of *Clematis* sp.
When it noticed that my camera was close to it,
it bent its front feet to demonstrate to me, just like a person showing his biceps brachii, indicating that it was strong.
The crab spider is too small. In order to shoot bigger, I continued to shoot close to it.
It saw that its warning didn't work, so it turned over and prepared to run away.
The crab spider showed off a clown tattoo on its belly.

马戏小丑文身
Circus Clown Tattoo

马利筋在武汉东湖湿地的园林绿化中，
主要是作为引蝶植物加以利用。
马利筋的叶片和花是蝴蝶的幼虫——毛毛虫的食物，
花还是昆虫的重要蜜源。

Milweed (*Asclepias curasavica*) is planted in the Wuhan East Lake wetland for landscaping.
Milweed is mainly used as a butterfly guide plant.
The leaves and flowers of milweeds are also the food of the larvae caterpillars of butterflies,
and the flowers are also the important nectar plants of insects.

花样蛋挞
Egg Tarts like Flowers

一只满蟹蛛躲在一年蓬的花心里。

它们横向移动，十分像螃蟹，故称“蟹蛛”。

它们通常长期待在一朵花上狩猎，

蚂蚁、苍蝇、蜜蜂等小型昆虫都是它们的食物。

A crab spider (*Thomisus onustus*) hides in the flower heart of the annual fleabane (*Erigeron annuus*).

They move laterally, very much like crabs, so they are called “crab spider”.

They usually stay on a flower for a long time to hunt.

Small insects such as ants, flies and bees are their food.

在此等候
Right Here Waiting

武汉东湖湿地的一个花园种植了一些虞美人。

雨后天晴，一只蜗牛爬上了虞美人种子球的顶部。

蜗牛站在顶端停了下来，回头看着身后的风景。

也许蜗牛觉得站在高处很孤独，它在寻找可以下去的路。

生活中，当我们达到某一个高度时，也会像蜗牛一样瞻前顾后。

Some cron poppies (*Papaver rhoeas*) are planted in a garden in Wuhan East Lake wetland.

After the rain, a snail climbed to the top of the cron poppy's seed ball.

The snail stops at the top and looks back at the scenery behind him.

Maybe the snail feels lonely standing high. It is looking for a way to go down.

In life, when we reach a certain height, we will look forward and backward like the snail.

找个台阶下
Find a Step Down

蜻蜓和野荷是武汉东湖湿地最常见的物种。
一只黄蜻蜓停在卷曲的荷叶上休息。
夕阳照在蜻蜓的头部，
蜻蜓身后的树林由于没有阳光的照射变成黑色。
蜻蜓好像在做秀爱心的动作，卷曲的荷叶就像一颗爱心。
蜻蜓似乎在告诉世界：
只要大家都献出一份爱心，世界将变得更加美好。

Dragonflies and wild lotuses are the most common species in Wuhan East Lake wetland.
A yellow dragonfly rests on the curly lotus leaf.
The setting sun shines on the head of the dragonfly,
and the woods behind the dragonfly turn black because there is no sunlight.
The dragonfly seems to be showing love heart, and the lotus leaf without stretching is like a heart.
Dragonfly seems to be telling the world:
as long as everyone gives a love, the world will become a better place.

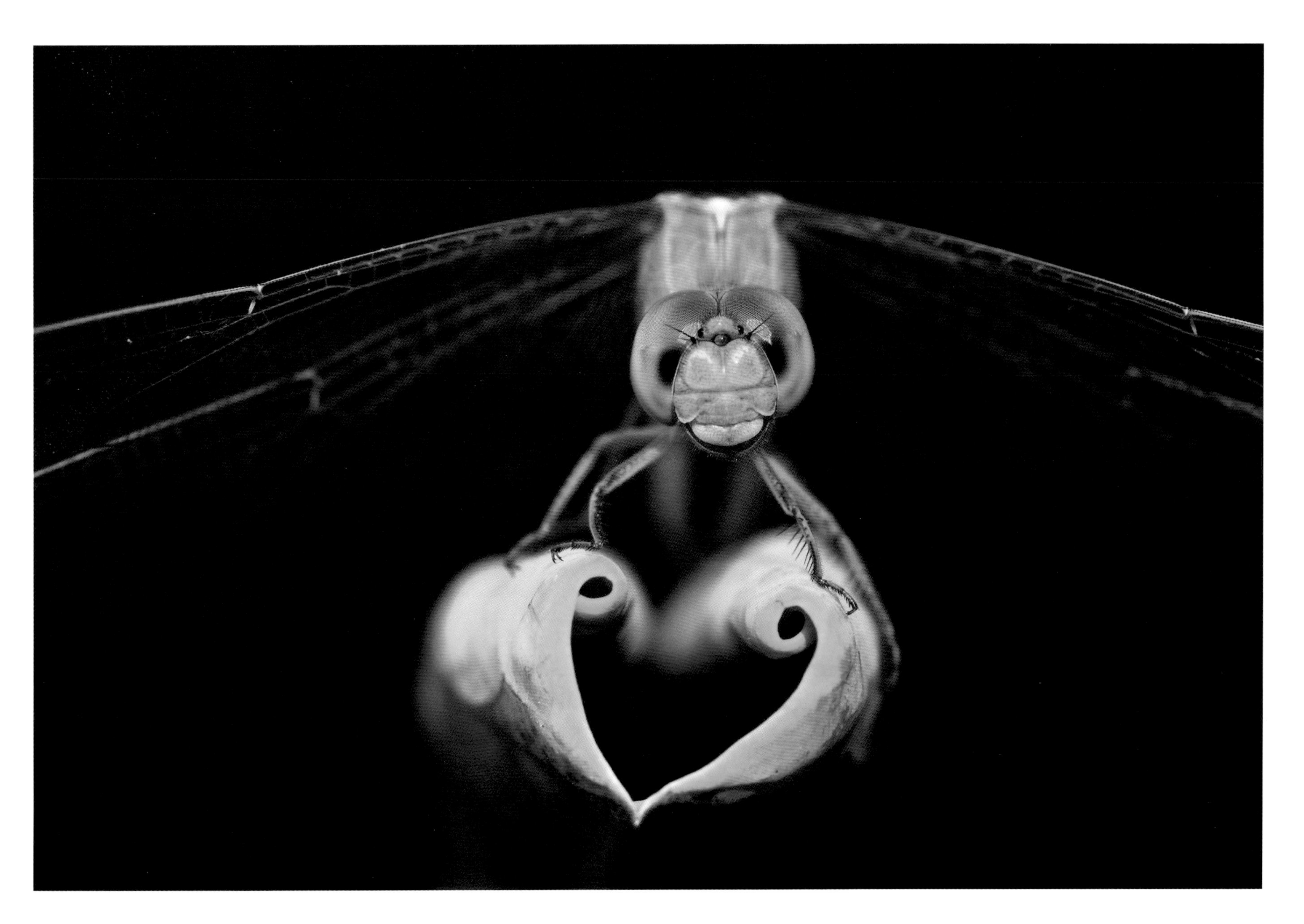

蜻蜓秀爱心
A Dragonfly Shows Love Heart

一只蓝眼睛的锥腹蜻停留在小莲蓬头上。

前面的荷叶遮挡住下面杂乱的环境，

我选择低的位置并用大光圈拍摄。

在梦幻的绿色中有一对锥腹蜻的蓝眼睛。

小莲蓬头的莲子就像一对绿眼睛，绿眼睛和蓝眼睛此刻一起看着我。

A male blue-eye Asian pintail (*Acisoma panorpoides*) stays on the seedpot of the lotus.
The lotus in front blocked the messy environment below.
I took a low position and shot with a large aperture.
In the dreamy green, a pair of Asian pintail lotus seeds in the seedpot are like a pair of green eyes.
Green eyes and blue eyes look at me together at the moment.

蓝眼睛和绿眼睛
Blue and Green Eyes

武汉东湖湿地是一个孕育神奇的地方。

我想给在荷叶边的一只剑角蝗拍张正面照。

我跪在地上慢慢接近，尽量不打扰它。

在拍摄时突然有了意外的发现——画面中出现了另一只不同颜色的剑角蝗。

四只小眼睛看着我，我觉得画面很有戏剧性。

眼睛里有好奇和惊讶，仿佛我才是这个星球的外来生物。

Wuhan East Lake wetland is a magical place.

In autumn, I wanted to take a front photo of a long-headed locust (*Acrida cinerea*) near the lotus leaf.

I knelt on the ground and approached slowly, trying not to disturb it.

When shooting, I suddenly made an unexpected discovery—another long-headed locust in different color appeared in the picture.

Four small eyes look at me, I thought the picture was very dramatic.

There was curiosity and surprise in four small eyes, as if I were an alien creature on this planet.

小四眼
Little Four Eyes

一群蝽宝宝出生后在卵壳旁边团聚，

它们在等待最后一个未出壳的兄弟。

蝽宝宝们有共同的身世，有着亲兄弟般的情谊。

接下来的几天，它们会像一个团队一样待在一起经历风雨。

蝽宝宝们会蜕一次皮，体色变成与它们母亲一样的黑灰色，

这样不容易被其他掠食者发现。

A group of stinkbug babies are reunited next to the egg shell after birth.

They are waiting for the last brother who hasn’t come out of the shell.

Stinkbug babies have a common life experience and brotherly friendship.

In the next few days, they will stay together like a team and experience the wind and rain.

Baby stinkbugs shed their skin once and become as black and gray as their mother,

so that they can’t be found by other predators easily.

兄弟连
Band of Brothers

在武汉东湖湿地的春天，

这种喜欢把蚜虫聚集在一起的放牧蚂蚁是靠蚜虫屁股后分泌的蜜露生活的。

蚜虫靠吸食植物的汁液生活。

它们的粪便是亮晶晶的液体，

含有丰富的糖，我们称之为“蜜露”。

蚂蚁非常爱吃蜜露，常用触角拍打蚜虫的背部，促使蚜虫分泌蜜露。

人们把蚂蚁的这一动作叫作“挤奶”，而把蚜虫比作蚂蚁的“奶牛”。

In the spring of Wuhan East Lake wetland,
this kind of grazing ant who likes to gather aphids lives on the honeydew secreted by the ass of aphids.
Aphids live on the sap of plants.
Their feces are shiny liquid and rich in sugar, which we call “honeydew”.
Ants love honeydew very much.
They often beat the back of aphids with their antennae to promote aphids to secrete honeydew.
People call this action of ants “milking”, and compare aphids to ants’ “cows”.

放牧
Grazing

一只中华大刀螳若虫站在白荷花花心里，
它在荷花花蕊里捉了一只小甲虫正在吃，
当我靠近它，准备给它来一个特写的时候，
它扭头用警惕的眼神看着我，
好像在对我说："别过来，这是我的午餐。"
我在多次的拍摄中发现，螳螂都会利用花朵的吸引力来等待捕捉采花粉和花蜜的昆虫。
这是它们的生存之道。

A nymph of a chinese mantis (*Tenodera sinensis*) stands in the white lotus.
It caught a little beetle in the lotus stamens and was eating.
When I approached it, the little mantis turned its head and looked at me with vigilant eyes,
as if saying to me, "don't come here, this is my lunch."
I have found in many of my shots,
the mantis uses the attraction of flowers to wait for catching pollen-gathering insects.
This is their way of survival.

小螳螂的午餐
Lunch of the Little Mantis

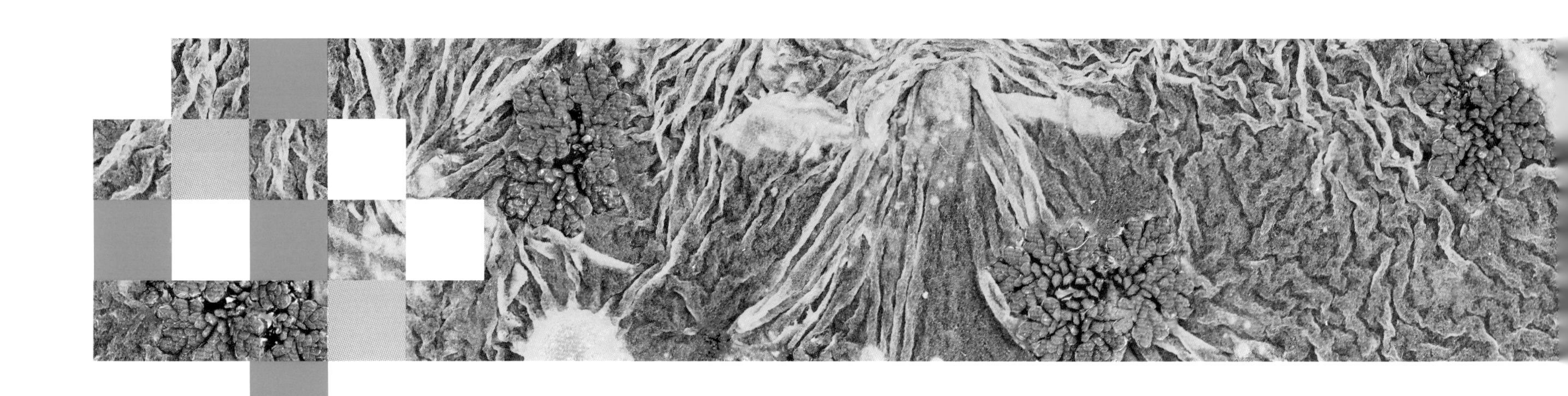

植物礼赞

In Praise of Plants

植物总会让我心情愉悦。有时，挂在墙壁上的一幅植物影像也会让我们驻足欣赏。植物缓解了我们生活和工作中的压力。植物世界仿佛联系着我们人类的精神世界，它们的特质指引着生命的归宿。

Seeing plants always makes me feel happy. Sometimes, an image of a plant hanging on the wall will let us stop and enjoy it. Plants relieve the pressure in our life and work. The plant world seems to be connected with our human spiritual world, and their characteristics guide the end result of life.

早春的池塘水面铺了一层薄膜样的藻类。
其间有一些满江红在生长，
红色与绿色交互结合。
春天的武汉是一个浪漫的城市，
东湖湿地总会有不同的色彩和情感。

The surface of the pond in early spring is covered with thin-film algae.
Occasionally, some red duckweed are growing,
and red and green are combined alternately.
Wuhan is a romantic city in spring.
East Lake wetland always has different colors and emotions.

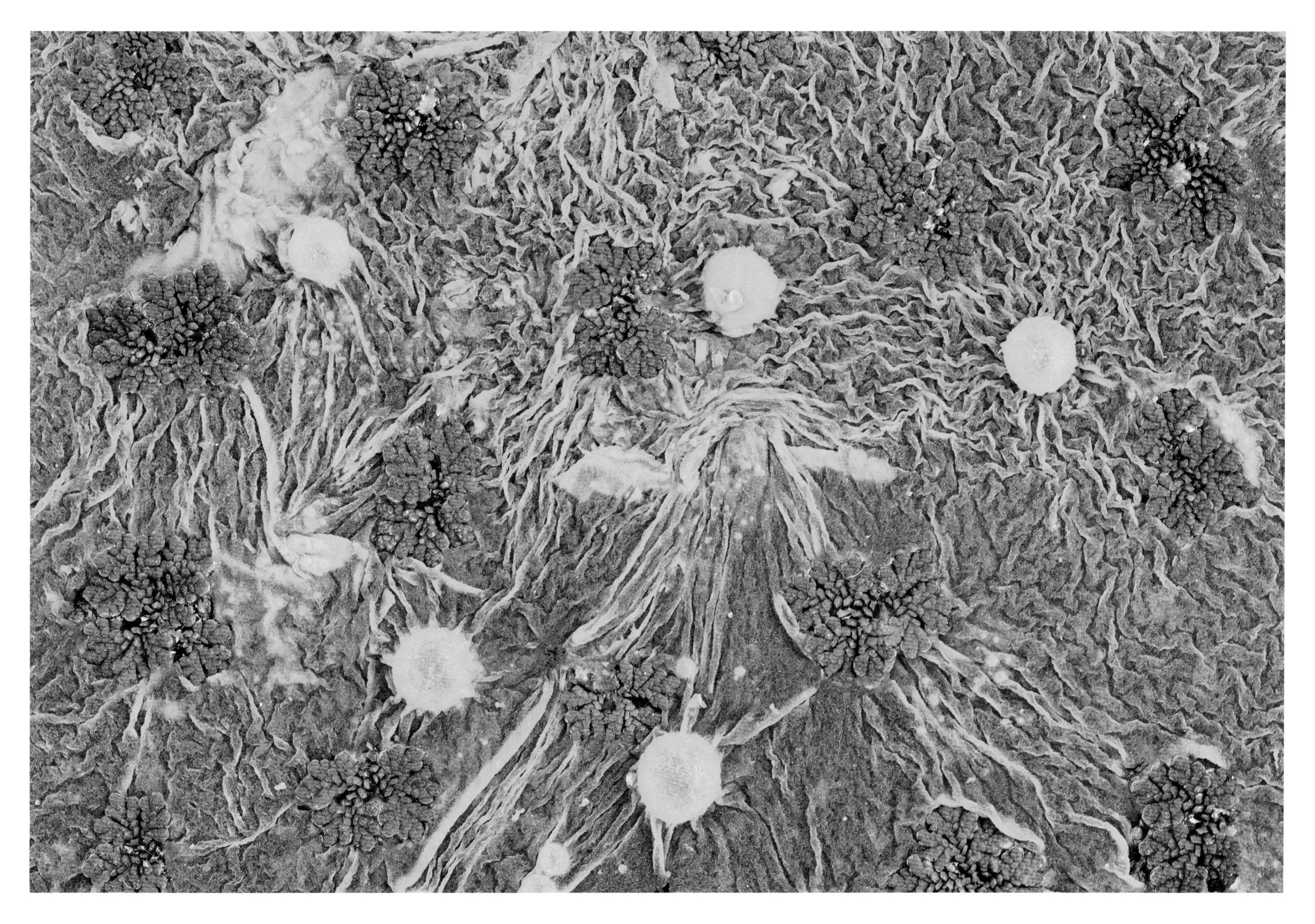

错开的春天
Stagger of Spring

爬山虎抛出圆盘状吸盘，顽强地促进其生长，延长其生存的时间。

在武汉郊区的树林里，我看到一个像粘在树皮上的手一样的黏性吸盘。

这瘦瘦的小手看起来充满进步的力量。

小手表明生命的脆弱和坚韧，有决心，不屈服于命运。

The Boston ivy (*Parthenocissus tricuspidata*) throws out a disc-shaped suction cup and makes tenacious efforts to promote its growth and prolong its survival time.

In the woods on the outskirts of Wuhan, I saw a sticky sucker like a hand stuck to the bark.

This thin little hand looks like the power of progress.

Small hands show the fragility and tenacity of life and the determination not to yield to fate.

坚持

Holding on

我在雨后的池塘边发现了这两片荷叶。

老旧和新生的荷叶让我联想到一位成人牵着一位孩子的手。

新生的荷叶就像是从老旧荷叶身上分出去的一部分。

这让我想到母亲在生下孩子以后渐渐老去的情形，

生命就是如此轮回的。

I found these two lotus leaves by the pond after the rain.

The old and new lotus leaves remind me of an adult holding a child's hand.

The new lotus leaf is like a part separated from the old lotus leaf.

This reminds me of the situation that the mother gradually grows old after giving birth to a child.

Life is like this reincarnation.

生生不息
The Circle of Life

武汉东湖湿地生长着各种颜色的睡莲。
太阳下山后，白睡莲闭合了花瓣。
白睡莲花就像一个睡美人在枕着睡莲叶片睡觉，
她在梦中低语着白天看到的一切。
睡莲叶的茎在白睡莲花周围形成了优美的线条，
同时也烘托了现场温馨柔美的氛围。

Water lilies in various colors grow in Wuhan East Lake wetland.
After sunset, a white water lily closed its petals.
The white water lotus is like a sleeping beauty sleeping on the lotus leaf.
She whispers everything that she sees during the day in her dream.
The stems of water lily leaves form beautiful lines around the white sleeping lotus,
and also set off the warm and soft atmosphere of the scene.

梦中的呢喃
Whispering in Dream

苹（俗称四叶苹）叶片的大小和颜色显示了它们在生长过程的四个阶段，

就好像苹生长时经历的四个季节。

水下的藻类和新生的苹，其叶片浸润在水中，

湿地见证了它们生长的过程和美丽。

The size and color of leaves of the four-leaved duckweed (*Marsilea quadrifolia*) show
the four stages of their growth process,
just like four seasons experienced by the four-leaved duckweed when growing.
Underwater algae and newborn leaves of the four-leaved duckweed infiltrate the water,
and the wetland witnesses their growth process and beauty.

水中四季
Four Seasons in Water

早春，水杉还没有长出新叶，

但池塘里已经错落有致地漂满了苹。

水杉的倒影映在池塘里，

我觉得这是自然艺术中一幅充满诗意的画面。

回家后，我把照片旋转了180度，

漂浮在水面的苹似乎飘飞在空中。

In early spring, the dawn redwood (*Metasequoia glyptostroboides*) has not grown new leaves,

but the pond has been dotted with four-leaved duckweed.

Reflections of the dawn redwood were in the pond,

which I thought was a poetic picture in natural art.

After returning home, I rotated the photo 180 degrees,

and the four-leaved duckweed floating on the water seemed to float in the air.

倒影的反思
Reflection on Reflection

中国鄂西的女孩喜欢在结香枝上打结，
祈祷她们能找到自己梦中的白马王子。
所以结香花也被称为梦花。
这两朵结香花就像一对经历了困难曲折后重逢的恋人。
“许愿结”是一个祈祷，它就像一个中国结，蕴含着一个美丽的故事，
它期待梦想可以成真！

Girls in western Hubei Province like to tie knots on branches of the oriental paperbush (*Edgeworthia chrysantha*),
praying that they can find their dream lover.
So the oriental paperbush's flower is also called dream flower.
These two oriental paperbush's flowers are like a pair of lovers who meet again after experiencing difficulties and twists.
"Wishing knot" is a prayer. It is like a Chinese knot. It contains a beautiful story.
It looks forward to the dream coming true!

经历曲折后的重逢
Reunion after Twists and Turns

武汉东湖湿地边的狐尾藻吸引了我的注意。

从上往下看，

许多正在生长的狐尾藻形成了一个平面，

但我觉得这有些单调。

我想表现一种梦幻的景象，

换个观察角度看到狐尾藻叶片之间形成了轮状的结构，

像盛开的绿色的花。

The foxtail algae (*Myriophyllum verticillatum*) beside Wuhan East Lake wetland attracted my attention.

From top to bottom, many growing foxtail algae formed a plane composition,

but I though it was a little monotonous.

I wanted to show a dreamy scene.

After changing the observation angle,

I saw a wheel-like structure formed between leaves of the foxtail algae,

which looked like blooming green flowers.

夏季的喜悦
Joy in Summer

武汉东湖湿地草地上的一朵虞美人花开始凋落，
看上去就像一个穿着红长袍、戴着帽子的小公主。
盛开的花儿是美丽的，凋落的花儿却是最有个性的，
因为它看尽繁华后知道世界是自己的。

A corn poppy (*Papaver rhoeas*) begins to wither on the grassland of Wuhan East Lake wetland.
It looks like a little princess in a red robe and hat.
Blooming flowers are beautiful, but fading flowers are the most personalized,
because they know that the world is their own after seeing the prosperity.

小公主
Little Princess

这朵凋零的莲花的形状让我想起一位穿着贵族服装的女士。
随着年龄的增长，最美丽的行为之一就是临终时无畏。
这确实是所有灵性的生命具有的一种最高尚的精神。

The shape of this withered lotus reminds me of a lady in aristocratic clothes.
As you grow older, one of the most beautiful behaviors is to be fearless on your deathbed.
This is indeed the most noblest spirit of all spiritual life.

女士
Lady

武汉植物园里生长着一片黄花羊蹄甲。
一大一小的两片羊蹄甲叶片重叠在一起，
逆光之下就像是对半切开的青苹果。
通过它们在形态上的相加作用，
自然界中的一些视觉元素相互组合会形成一个新的视觉形象。

St. Tomas trees (*Bauhinia tomentosa*) grow in Wuhan Botanical Garden.
Two leaves of the tree, one large and one small, overlapped together.
Under the backlight, it was like a green apple cut in half.
Through their additive action in form,
some visual elements in nature will combine with each other to form a new visual image.

青苹果
Green Apple

武汉东湖湿地生长着很多的金银莲花。
我发现两朵金银莲花依偎在一起的倒影能带来相反的视觉。
利用水面的反光可以让我看到金银莲花的底部，
这是我无法在水下往上拍摄时获得的角度。

There are many water snowflakes (*Nymphoides indica*) in Wuhan East Lake wetland.
I found that the reflection of two water snowflakes snuggling together could bring opposite vision.
Using the reflection of the water surface, I could see roots of the water snowflakes,
which was the angle I couldn’t get when shooting up underwater.

水中的依偎
Snuggling in the Water

灯笼象征着幸福、光明、活力、圆满与富贵。

从中国古代到现在，人们一直都有过新年挂灯笼的习俗。

同时，这也是营造喜庆氛围的手段。

我看到金铃花（俗称灯笼花）正在盛开，

大自然中开放的一对小灯笼花给我带来新春的吉祥。

Lanterns symbolize happiness, brightness, vitality, perfection and wealth.

From ancient China to now, people have always had the custom of hanging lanterns for a new year.

At the same time, it is also a means to create a festive atmosphere.

I saw the Chinese lanterns (*Abutilon pictum*) in full bloom.

A pair of small chinese lanterns open in nature bring me the auspiciousness of the Spring Festival.

新春灯笼

Lanterns for the Spring Festival

风与光的季节

Season of Wind and Light

春天的风让人感叹生命的开始；夏天的太阳让人感叹生命的严酷；秋天的色彩让人感叹生命的美好；冬天的光让人感叹生命的落寞。时光流逝在风与光中，我们静静感受自然的变化和生命的成长。

The spring wind makes people sigh the beginning of life; the summer sun makes people sigh the harshness of life; the color of autumn makes people sigh the beauty of life; the light of winter makes people sigh the loneliness of life. Time passes in the wind and light, we quietly feel the changes of nature and the growth of life.

在武汉东湖湿地，一只大飞蛾从灌木丛中飞向了一棵大树。

当我走过去的时候，却没有找到飞蛾的踪影。

我细看眼前的树皮发现，飞蛾平摊开的翅膀完美地融入树皮的纹理中。

这只飞蛾的身体保护色是它躲避天敌的方式。

In Wuhan East Lake wetland, a big moth was flying from the bush to a big tree.

When I walked past, I didn’t find the trace of the moth.

After looking at the bark carefully, I found that the moth’s spread wings perfectly integrated into the texture of the bark.

The moth’s body protective color is its way to avoid natural enemies.

树皮上的伪装
Camouflage on Bark

武汉东湖湿地种植了很多种类的竹子。

在中国的传统文化中，竹子是刚正不阿的形象，也象征顽强的生命。

我听到竹叶在风中“沙沙”作响，就像在窃窃私语。

我使用两次曝光拍摄风中的竹子，让竹子有中国画的韵味。

Many kinds of bamboo are planted in Wuhan East Lake wetland.

In Chinese traditional culture, bamboo is an image of integrity, and it also symbolizes tenacious life.

I heard the rustle of bamboo leaves, which seemed to whisper in the wind.

I used two exposures to shoot bamboo in the wind, so that bamboo has the charm of Chinese painting.

竹的呢喃
Whispering in the Wind

在湖北大别山溪流边的草上，

每隔三四米就有一只金属色的透顶单脉色蟌停留。

透顶单脉色蟌有着很强的领地意识。

如果有其他雄色蟌飞过，停留在草上的雄色蟌会飞起驱赶它们。

我穿着连体防水服在冰冷的溪流中泡了两个多小时，直到能接近它们。

一阵微风吹过，雄色蟌在空气中感到一丝紧张的气氛，

它展开了翅膀，姿态高贵而优雅，就像在风中绽放的一朵神秘之花。

On the grass beside the stream in Dabie Mountain, Hubei Province,

a metal colored single-veined pierogi stops every three or four meters.

The single-veined pierogi has a strong sense of territory.

If other males fly by, the male single-veined pierogi staying on the grass will fly to drive away them.

I wore a one-piece waterproof suit and soaked in the cold stream for more than two hours until I could get close to them.

A breeze blew, a male single-veined pierogi felt a slight tension in the air, and it spread its wings.

Its posture is noble and elegant, just like a mysterious flower blooming in the wind.

溪流的微风
Breeze in the Stream

在武汉东湖湿地的冬季，干枯的藤蔓进入了休眠状态。

春天来临，藤蔓开始发芽生长。

夏天，藤蔓沿着大树越爬越高。

秋天，藤蔓叶片开始从下往上变成红色。

藤蔓孤独地走过了季节，但它收获了成长的喜悦。

In the winter of Wuhan East Lake wetland, the dry vines entered a dormant state.
When spring comes, vines begin to sprout and grow.
In summer, the vines climb higher and higher along the big tree.
In autumn, the vine leaves begin to turn red from bottom to top.
The vine has gone through the season alone, but it has harvested the joy of growth.

走过季节
Through the Season

秋天，高大的三球悬铃木（俗称法国梧桐）的树皮呈现出斑驳的颜色。

泛黄的树叶在秋风中簌簌飘落。

二次曝光让眼前的场景更加梦幻。

红色、棕色、乳黄色和灰色的过渡，为这张照片增添了油画般的效果。

In the autumn of Wuhan East Lake wetland, the bark of the tall France plane tree (*Platanus orientalis*) presents mottled color.

Yellow leaves rustle and fall in the autumn wind.

The second exposure makes the scene more dreamy.

The transition of red, brown, cream and gray adds an oil painting effect to this photo.

秋日印象
Autumn Impression

武汉东湖湿地边的树林里，蜘蛛网在阳光的照射下变幻出多种色彩。

多数时候，蜘蛛网就像迷人的彩虹灯丝，片刻点亮暗色的树林间。

一只小蜘蛛在它编织的丝网上捉到了一只小昆虫，此时它正在享用美食。

蜘蛛就像魔术师在舞台中央进行表演。

In the woods near Wuhan East Lake wetland, spider webs change a variety of colors under the sunlight.

Most of the time, the spider web is like a charming rainbow filament, which lights up the dark woods for a moment.

A small spider caught a small insect on its woven silk screen.

It is enjoying delicious food. Spiders act like magicians in the middle of the stage.

光影魔术
Light and Shadow Magic

武汉东湖湿地的公园里，春天的落日下，

含苞待放的虞美人在拥抱。

花儿的姿态像两个恋人，有着属于它们的爱情。

明天的阳光下，它们也许会绽放，继续拥抱太阳！

In Wuhan East Lake wetland, budding corn poppies are embracing under the setting sun in spring.

The posture of flowers is like two lovers. Flowers have their love.

In tomorrow's sunshine, budding corn poppies may bloom.

They will continue to embrace tomorrow's sun!

阳光下的拥抱
Embrace in the Sun

武汉东湖湿地的冬天，

鸡爪槭的断枝横落在水岸边。

黄昏时，镜头里闪烁迷离的波光就像舞厅的灯光。

鸡爪槭种子仿佛一对对恋人在浪漫的灯光下跳贴面舞。

In the winter of Wuhan East Lake wetland,

broken branches of the Japanese maple (*Acer palmatum*) fall horizontally on the Bank of the Lake.

At dusk, the flickering wave light in the lens is like the light of a dance hall.

The Japanese maple's seeds are like a pair of lovers dancing in the romantic light.

贴面舞
Cheek-to-cheek Dancing

伪蝎体形很小，一般体长不超过8毫米。

生活在落叶层、土壤中、树皮和石块下。

我发现一只伪蝎躲藏在卷曲的枯叶中，

逆光下枯叶的色彩像是被火燃烧了一样。

我觉得伪蝎更像一只龙虾。

但是它不是在一个熔炉里等待被烹饪，而是从热的光源里获得了生命能量。

The Pseudoscorpion (Pseudoscorpionida) is very small and generally no longer than 8 millimeters.

It lives in the deciduous layer, soil, bark and stones.

I found a pseudoscorpion hiding in curly dead leaves.

The color of the dead leaves in the backlight seems to be burned by fire.

I think pseudoscorpion more like a lobster.

But instead of waiting to be cooked in a furnace, it gets life energy from a hot light source.

热能中的伪蝎
Pseudoscorpion in Thermal Energy

在武汉东湖湿地附近，鸡爪槭的叶片落在草地上，

早上的霜覆盖在它们上面。

叶落归根，总会带来离别的伤感情绪。

四季轮回，我从这种回归中看到的是明年的希望。

冬天来了，春天还会远吗?

Near Wuhan East Lake wetland,

the Japanese maple's leaves fall on the grass and are covered with frost in the morning.

When leaves fall to their roots, they will always bring the sadness of parting.

The four seasons reincarnate. What I see from this return is the hope of the coming year.

If winter comes, will spring be far behind?

冬日
Winter

盛夏的夕阳中，一只螽斯倒挂在狗尾巴草上。
太阳完全落山后，属于螽斯的夜生活才真正开始。
雄螽斯通过发出自己独特的鸣声寻找配偶，
吸引同种雌虫前来交配。

In the setting sun of midsummer, a katydid (Tettigoniidae) hangs upside down on dog's tail grass (*Setaria viridis*).
After the sun sets completely, the nightlife of katydids really begins.
Male katydids find mates by making their own unique song,
attracting females of the same species to mate.

日落的前奏
Prelude to Sunset

呢喃与夜曲
Whisper and Nocturne

夜晚是生命的另一种绽放，呢喃是生物之间的情话。不是所有的语言我们都能懂，但自然发出的声音就像音乐一般给我们启迪。不是所有乐曲都能带给我们能量，但一个真诚的微笑，就能让我们温暖。

Night is another bloom of life, and whispering is the love talk between creatures. Not all languages are intelligible to us, but the natural sounds give us enlightenment like music. Not all music can bring us energy, but a sincere smile can bring us warmth.

湿地的夜晚是各种鸣虫唱歌的时刻。
晚间，两只螽斯在草上轻声鸣叫，
它们通过声音在感知同类的存在，
这是它们彼此联系的方式。
螽斯颜色和草叶颜色融为一体，
它们仿佛在夜色里吟唱了一支小夜曲。

The night in the wetland is a time for all kinds of singing insects.
At night, two katydids chirp softly on the grass.
They perceive the existence of their own kind through sound,
which is the way they contact each other.
The color of katydids and the color of grass leaves are integrated.
They seem to sing a serenade at night.

小夜曲
Serenade

在武汉市黄陂区的木兰湖边，
一只蝗虫正在灌木丛中蜕皮。
就像童话《皇帝的新装》中的皇帝，
它脱下了一件衣服，最后什么也没穿。

By Mulan Lake in Huangpi District of Wuhan.
A locust is molting in the bush.
As the emperor in the fairy tale *The Emperor's New Clothes*,
it took off a dress and didn't wear anything at last.

蝗虫的新装
New Clothes of Locusts

晚上夜拍时，我在武汉植物园的睡莲池边发现了这个场景。

一只水虿（蜻蜓的幼虫）爬上睡莲叶片，脱壳羽化。

青蛙对迅速移动的昆虫会做出迅速反应，

但对眼前正在慢慢羽化的蜻蜓却视而不见。

水虿生活在水中时吃一些水里的小昆虫、小鱼和蝌蚪。

蝌蚪长大变成青蛙后却可以吃蜻蜓。

这片睡莲叶片就像一个生命轮换角色的舞台。

When shooting at night, I found this scene by the water lily pool in Wuhan Botanical Garden.

A nymph of the dragonfly climbs on the water lily leaf and is eclosing.

Frogs react quickly to fast-moving insects,

but turn a blind eye to the dragonfly that is slowly eclosing.

When live in the water, nymph of the dragonflies eat some small insects, fish and tadpoles.

When tadpoles grow up and become frogs, they can eat dragonflies.

This water lily leaf is like a stage of life rotation.

生命的舞台
Stage of Life

美人蕉的叶背底下，

从一个蝶蛹里钻出了一只二尾蛱蝶。

二尾蛱蝶在等待自己的翅膀长得更大、更强壮，

它准备去迎接未来的考验。

一切才刚刚开始。

Under the leaf back of a canna (*Canna indica*), a *polyura narcaea* emerged from a chrysalis.

The Polyura narcaea is waiting for its wings to grow bigger and stronger.

It is ready to meet the test of the future.

Everything has just begun.

蜕变
Change Qualitatively

夏夜，武汉东湖湿地旁边的树林里有很多蝉的幼虫会从土里钻出来羽化。
我在一处古老爬藤上发现密布着很多蝉蜕。
我以天空为背景，逆光下蝉蜕就像一只只金色的生命在爬藤上攀登。

On the summer night, there are many cicadas' larvae in the woods next to
the Wuhan East Lake wetland, which will drill out of the soil and emerge.
I found a lot of cicadas' molting shells on an ancient climbing vine.
I took the sky as the background. In the backlight, these cicadas' molthing shells
were like golden lives climbing on the vine.

极限攀登
Extreme Climbing

红锯蛱蝶身上的图形纹似美国影星玛丽莲•梦露的嘴形，
前翅就像梦露美丽的肩形，
人们为了纪念她，故又把此蝶命名为“梦露蝶”。
两只梦露蝶在一根竹枝上交尾，
它们侧面翅膀的左右对应姿态像一只梦露蝶在展开翅膀。

The graphic pattern on the red lacewing (*Cethosia Biblis*) is similar to the mouth shape of American film star Marilyn Monroe.
The front wing of the red lacewing is like the beautiful shoulder shape of Monroe.
In order to commemorate her, people named the red lacewing Monroe butterfly.
Two red lacewings mate on a bamboo branch, and the corresponding posture of the wings on the sides of the two cethosia biblis is like a red lacewing spreading its wings.

合二为一
Two Becomes One

在武汉东湖湿地，晚上的枯叶蛱蝶倒挂在枯树枝上休息。

枯叶蛱蝶前翅顶角和后翅臀角向前后延伸，

呈叶柄和叶尖形状。

具有拟态行为的枯叶蛱蝶和周围的枯叶融为一体。

In Wuhan East Lake wetland,

a dead-leaf butterfly (*kallima inachus*) hangs upside down on the dead branch to rest at night.

The dead-leaf butterfly front wing apex angle and rear wing hip angle extend forward and backward in the shape of petiole and tip.

The dead-leaf butterfly with mimicry behavior is integrated with the surrounding dead leaves.

枯叶模仿秀
Imitation Show of Dead Leaves

在武汉东湖湿地的一次夜晚观察时，

我发现一对黑脉金斑蝶在一朵菊花的茎秆上交尾。

它们在春天离开过冬地时交配。

黑脉金斑蝶是迁徙性蝴蝶。

少数黑脉金斑蝶在迁徙过程中会出现在武汉的湿地。

During a night observation in Wuhan East Lake wetland,

I found a pair of monarch butterfly (*Danaus plexippus*) mating on the stem of a chrysanthemum.

They mate when they leave the wintering land in the spring.

Monarch butterfly is a migratory butterfly.

A few monarch butterfly will appear in the wetlands of Wuhan during migration.

爱之花
Flower of Love

盛夏的中午，

一只豆娘的幼虫爬上蓝睡莲的花心后正在蜕变为成虫。

没有什么能阻挡它对飞上蓝天的向往。

水孕育着生命，水让睡莲和豆娘都得到重生。

At noon in midsummer,

a damselfly larva is transforming into an adult after climbing the flower center of the blue water lily.

Nothing can stop it from flying into the blue sky.

Water breeds life. Water gives birth to both water lily and damsefly.

睡莲里的重生
Rebirth in the Water Lily

生命继续

Life Goes On

当母亲生下我们时，我们就开始了人生的旅程。年轻时，我们在挫折中奋斗，在失败中追求胜利。想要学习，却永无止境。我们庆幸终于找到了自我。多年后，我们已经苍老，却再也回不到母亲的怀抱。

When our mother gave birth to us, we began the journey of life. When we were young, we struggled in frustration and pursued victory in failure. Want to learn but never stop. We are glad to finally find ourselves. Many years later, we are old, but we can no longer return to our mother's arms.

武汉东湖湿地的夏季是多种植物和昆虫茁壮成长的季节。

田野里紫色的千屈菜花盛开，一对鹿蛾在田埂边的红花羊蹄甲的叶片上交尾。

红花羊蹄甲的叶片像一个青苹果，是鹿蛾的栖息乐园。

荒野因为这些小生命的存在，才变得更美。

The summer of Wuhan East Lake wetland is the season for the growth of a variety of plants and insects.
In the field, the purple loosestrife (*Lythrum salicaria*) is in full bloom,
and a pair of Ctenuchidae mate on a leaf of the mountain ebony (*Bauhinia blakeana*) on the edge of the field.
The leaf of the mountain ebony look like a green apple,
while the leaves of the mountain ebony are the habitat paradise of Ctenuchidae.
The wilderness is more beautiful because of the existence of these small lives.

青苹果乐园
Green Apple Paradise

武汉东湖荷花园的夏季，有些荷花已经凋谢，露出了小莲蓬。

一对灰蝶在小莲蓬的顶端交尾。

我蹲下来拍摄时，发现背景中的荷叶正好挡住了杂乱的环境。

我慢慢接近灰蝶，动作必须轻，避免打搅它们。

虽然它们显得很渺小，但在它们眼里，它们站在世界之巅。

In the summer of Wuhan East Lake Lotus Garden, some lotus flowers have withered,
revealing their seedpods.
A pair of gossamer-winger butterflies (Lycaenidae) mate at the top of the seedpod of a lotus.
When I squatted down to shoot,
I found that the lotus leaf behind the background just blocked the chaotic environment.
I approached the grey butterflies slowly.
I must move gossamer-winger to avoid disturbing them. Although they seem small,
they stand at the top of the world in their eyes.

世界之巅
Top of the World

一对螳螂在梭鱼草花上交尾，
前臂弯曲做祈祷状的螳螂此刻在祈祷爱的来临。
雄螳螂在交尾之后如果不及时跑掉，往往会成为雌螳螂的猎物。
这是自然界的生存法则。

A pair of mantises are mating on a flower of the pickerelweed (*Pontederia cordata*).
The mantis with its forearms bent for prayer is praying for the coming of love.
If the male mantis does not run away in time after mating, it will often become the prey of the female mantis, which is the survival law of nature.

爱的祈祷
Prayer of Love

我穿着连体防水服蹲伏在大别山的溪流边，

等待拍摄透顶单脉色蟌的交尾瞬间。

拍摄透顶单脉色蟌的交尾需要耐心和运气，

因为透顶单脉色蟌的交尾时间通常只有十几秒钟。

它们交尾的形态像一个爱心的造型。

I was wearing a one-piece waterproof suit and crouching by the stream of Dabie Mountain,

waiting for shooting the mating moment of single-veined pierogis.

Shooting the mating of single-veined pierogis requires patience and luck,

because the mating time of single-veined pierogis is usually only more than ten seconds.

They are in a form of love during mating.

爱心组合
Love Combination

春天里的小生命们都在向世界展示它们各自的美丽。
一对七星瓢虫在含苞待放的花蕾上交尾，
这是一个美好的预兆——它们的爱情已经萌发了。
生命如花朵般美丽！

The little creatures in spring are showing their beauty to the world.
A pair of ladybugs (*Coccinella septempunctata*) mating on budding flower buds is a good omen–their love has sprouted.
Life is beautiful as a flower!

爱的萌芽
Sprout of Love

当我发现这一对苏铁灰蝶在藤蔓上交尾时，
我觉得它们像是在接吻。
苏铁灰蝶又被称为双头灰蝶，
它们翅膀尾部的那个斑点像是一个义眼，
而它们翅膀尾部的突起又像是头部的触须。
柔和的光线照射在苏铁灰蝶身后的芭蕉叶上，
螺旋状的藤蔓和交尾的灰蝶之间仿佛有一种美好的寓意。

When I found this pair of *Chilades pandava* mating on the vine,
I felt they were kissing.
The *Chilades pandava* is also known as the double-headed grey butterfly.
The spots on the tails of their wings are like prosthetic eyes,
and the protrusions on the tails of their wings are like tentacles on the head.
The soft light shines on the banana leaves behind *Chilades pandava*.
There seems to be a beautiful implication between the spiral vine and the mating *Chilades pandava*.

螺旋的爱
Spiral Love

在湖北大别山的稻田边，

一对红脊长蝽在藤蔓的叶片上交尾。

心形的叶片像项链上的心形吊坠，

是红脊长蝽的爱情见证。

爱与生命在此刻串联在一起。

A pair of *Tropidothorax elegans* mate on the leaves of vines
near the rice field in Dabie Mountain, Hubei Province.
The heart-shaped leaf is the love witness of *Tropidothorax elegans*,
which is like a heart-shaped pendant on a necklace.
Love and life are connected at this moment.

心形吊坠
Heart Shaped Pendant

雄性棒络新妇蜘蛛比雌性棒络新妇蜘蛛个头小很多，

只有当雌性棒络新妇蛛用蜘蛛网捕捉到昆虫并进食时，

雄性棒络新妇蛛才有机会迅速接近雌性棒络新妇蛛进行交配。

一只不吃不喝的雄性棒络新妇蜘蛛一直跟在雌性棒络新妇蜘蛛的后面，

在网的另一边保持距离。

雄性棒络新妇蜘蛛知道如果贸然追求爱情可能会失去性命，

它在等待生命中一个最佳的机会，这个机会可能只有一次。

The male *Nephila clavata* is much smaller than the female.

When the female *Nephila clavata* catches insects with a spider web and eats,

the male *Nephila clavata* will seize the opportunity to quickly approach the female bridal spider for mating.

A male *Nephila clavata* who doesn’t eat or drink has been following the female.

It keeps a distance on the other side of the net.

The male *Nephila clavata* knows that he may lose his life if he rashly pursues love.

He is waiting for the best opportunity in his life, which may only be once.

危险的接触
Dangerous Contact

一对水黾在水中交配，

它们在水面一起划行，一起觅食。

水黾栖息于静水水面或溪流缓流水面上。

其身体细长，非常轻盈；

前脚短，可以用来捕捉猎物；

中脚和后脚很细长，长着具有油质的细毛，具有防水作用。

A pair of water skippers mate in the water.
They row together of the water and look for food together.
The water skipper inhabits the still water surface or the slow flow water surface of the stream.
Its body is slender and very light;
its forefoot is short and can be used to catch prey;
its feet are slender and long with oily fine hair,
which has the function of waterproof.

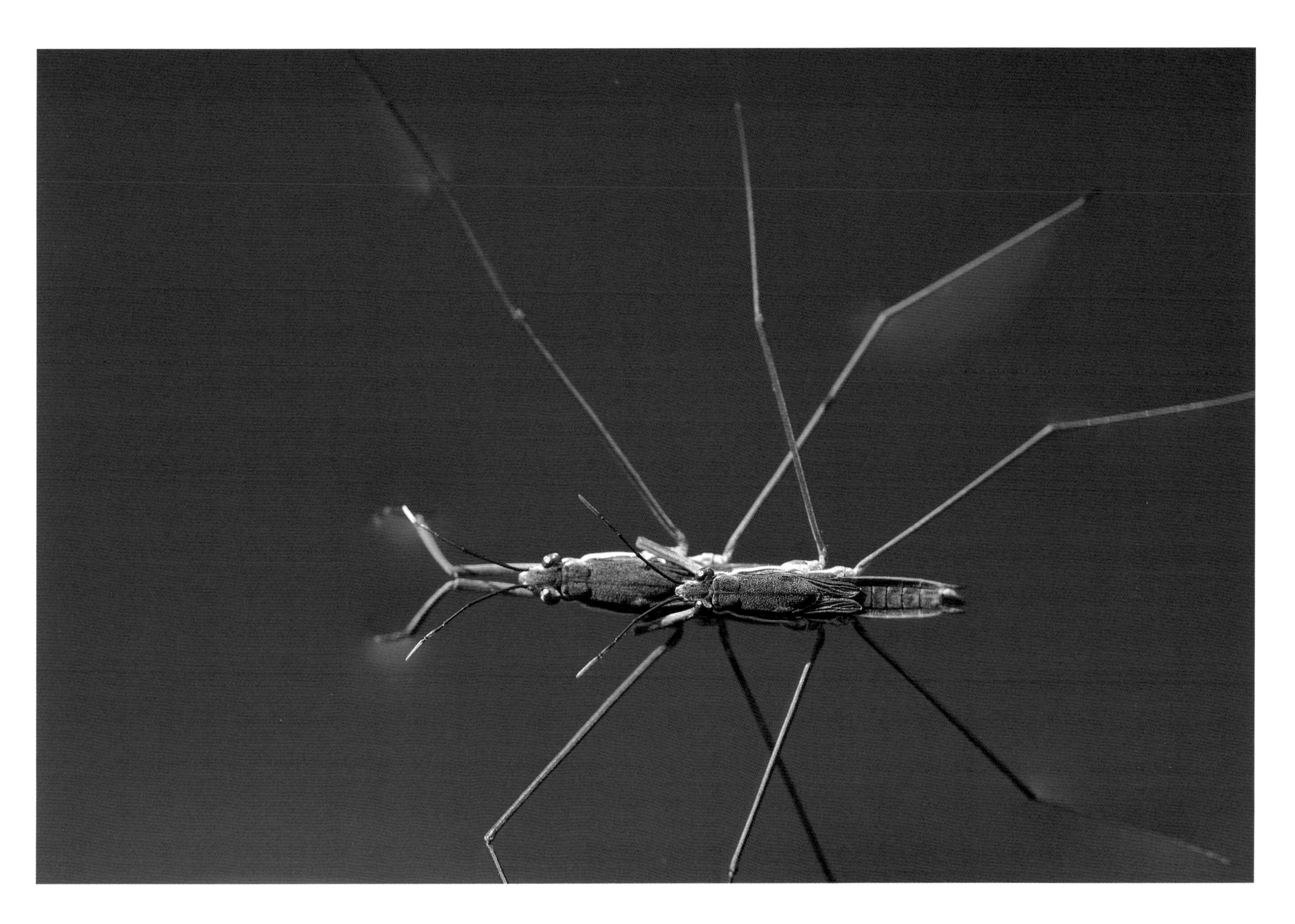

热情地拥抱
Romantic Contact

艺术与情感
Art and Emotion

艺术源于自然，自然中许多事物都是美丽的。在我们被美丽打动后，情绪也会变得激动。与自然亲密接触和交流的感受，让我们的创意永远新鲜。此刻，乐器不再是首选，相机和镜头是我们情感表达的工具。

Art comes from nature. Many things in the nature are beautiful. After we are moved by beauty, our emotions will also become excited. The feeling of close contact and communication with nature makes our creation forever fresh. At this moment, musical instruments are no longer the first choice. Cameras and lenses are tools for the expression of our emotions.

我用手机的微距模式拍摄还没有完全盛开的虞美人。

手机的镜头就像一个内窥镜，

让我看到一个奇妙的景象。

虞美人的花心就像一只精灵的眼睛。

我迷失在花的瞳孔中。

I use the macro mode of my mobile phone to shoot a cron poppy that is not yet in full bloom.
The len of the mobile phone is like an endoscope,
which makes me see a wonderful scene.
The interior of the cron poppy is like the eye of a flower elf.
I'm lost in the pupil of the flower.

花之瞳
Pupil of the Flower

在武汉植物园草坪上种植着一片太空瓜果。

清晨，一只七星瓢虫沿着藤蔓的线条走了几个来回。

它绕了几个圈圈后有点晕了，停下来休息。

我觉得瓢虫就像在玩过山车游戏。

人生也是有起有伏，就像瓢虫沿着藤蔓运动的轨迹。

Space melons and fruits are planted on the lawn of Wuhan Botanical Garden.

Early in the morning, a ladybug walked back and forth along the lines of the vine.

It was a little dizzy after several rounds.

It stopped to have a rest. I think the ladybug was playing a roller coaster game.

Life also has ups and downs,

which is like the trajectory of the ladybug moving along the vine.

生活就像过山车
Life Is like a Roller Coaster

雨后的湖北大别山，

太阳出来时气温开始升高，

这只华南雨蛙趴在农庄边的棕榈叶上等待它的伴侣。

我的拍摄让它有所警觉，它此刻的姿态看上去好像在跳独舞。

棕榈叶上的明暗变化像舞厅的射灯。

阳光使棕榈叶中心及其放射状复叶反映了这一自然过程。

In Dabie Mountain, Hubei Province, after rain, the temperature began to rise when the sun came out.

This South China rain frog (*Hyla simplex*) lied on the palm leaves beside the farm waiting for its partner.

My shooting made him alert.

It looked like he was dancing alone at the moment.

The change of light and shade on the palm leaves is like a spotlight in a dance hall.

Sunlight makes the center of palm leaves and its radial compound leaves reflect this natural process.

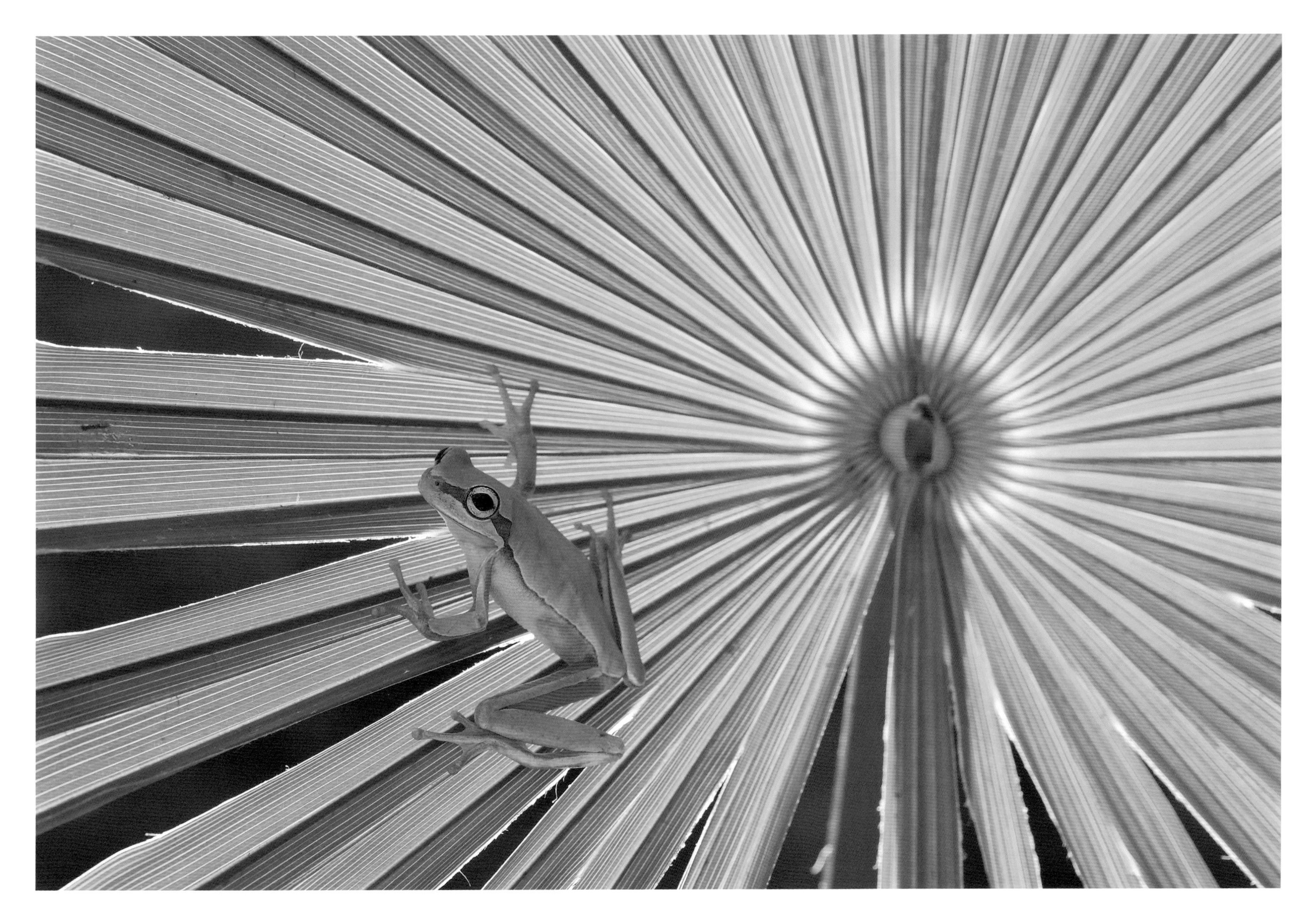

射灯下的独舞
Solo Dance under Spotlight

窗外在下雨，

一只豆娘飞到窗台上盆栽的野天胡荽（俗称铜钱草）下躲雨。

从侧面看去，豆娘就像一位优雅的女士在打伞。

我发现，以窗外的白墙为背景拍摄缺少了一些自然的色彩，

却可以让画面少一些杂乱。

在自然微距摄影中，

最大的惊喜就是你可以随时随地去发现一些趣味。

It was raining outside the window,
and a damselfly flew to the windowsill to hide under the potted penny grass (*Hydrocotyle vulgaris*).
From the side, the damselfly looks like a graceful lady holding an umbrella.
I found that shooting with the white wall outside the window as the background lacked some natural color,
but it made the picture less cluttered.
The biggest surprise in nature macro photography is that you can find something interesting anytime, anywhere.

打伞的豆娘

Damselfly with Umbrella

我喜欢在这棵龙枣树的枝干下观察，

因为这些枝条蜿蜒向上，奇形怪状，

给生命增加了很多的抽象和想象。

每年，武汉东湖湿地会吸引很多种鸟来越冬。

一个偶然的瞬间，一群迁徙的鸟从这些枝条上方飞过，然后消失在天空。

I like to observe under the branches of this jujube tree,

because these branches are winding upward and strange,

adding a lot of abstraction and imagination to life.

Every year, Wuhan East Lake wetland attracts many kinds of birds to overwinter.

In an accidental moment,

a group of migratory birds flew over these branches and disappeared into the sky.

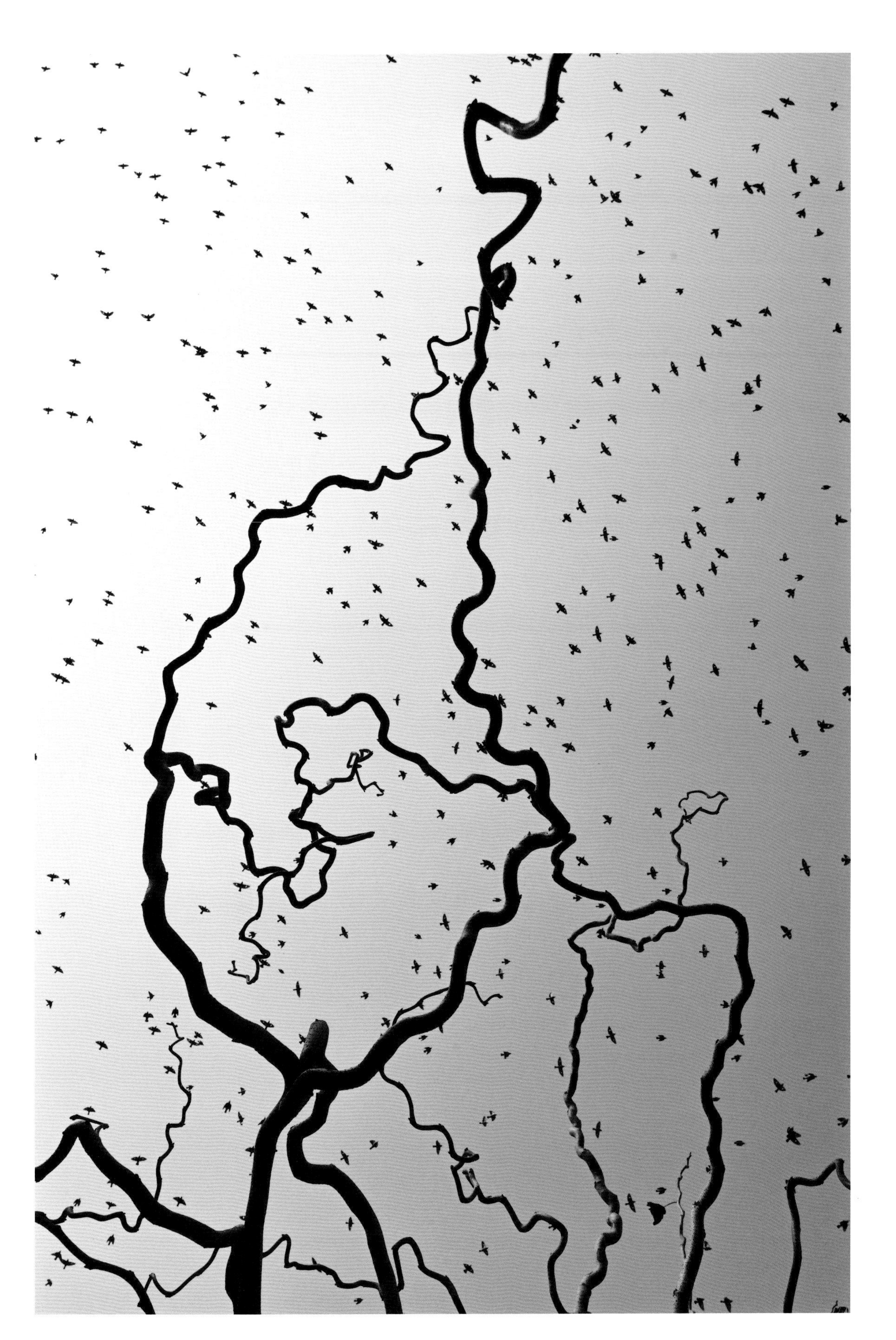

飞鸟集
Stray Birds

合欢花树枝上的一个螵蛸里钻出了很多小螳螂。

一只小螳螂站立在合欢花的丝状花蕊上。

这只小螳螂的前足沾上了合欢花的花粉，它正在清理它的前足。

画面中的小螳螂显得乖巧可爱，它像一个有思想的生命，躲在花蕊中看着我。

There are many small mantises in a cuttlebone on the branch of the silktree (*Albizia julibrissin*).
A small mantis stands on the filamentous stamens of a flower of the silktree.
The little mantis's front foot is stained with pollen from the flower.
It is cleaning its front foot.
The little mantis in the picture looks cute and clever.
It looks like a thoughtful life hiding in the stamens and looking at me.

思想者
Thinker

武汉是一个花城。

武汉东湖湿地公园种植了一些绣球花。

一个冬季过后，我发现干枯绣球花的花瓣脉络和叶脉都有着精致的细节。

干枯绣球花的花瓣叠加在叶脉上的组合就像蕾丝的花形结构。

自然之美也可以是经典的时尚艺术。

Wuhan is a city of flowers.

Some *Hydrangea macrophylla* are planted in Wuhan East Lake Wetland Park.

After a winter, I found that the petal veins and leaf veins of the dried *Hydrangea macrophylla* had exquisite details.

The combination of dried petals superimposed on the leaf veins is like the flower-shaped structure of lace.

The beauty of nature can also be a classic fashion art.

蕾丝
Lace

在春季，武汉每年都会出现一次快速降温。

清晨，不太活跃的豆娘停留在蒲公英上。

为了不惊飞它，我趴在地上悄悄靠近，想给豆娘来一个正面照。

此时豆娘举起一只前足擦眼睛，我用高速连拍记录了这一刻。

回家后，我从几张高速连拍的照片中挑选出这张。

那一刻，豆娘的前足停留在空中，就像一位演讲家即兴而为的一个手势，

它在发表那个著名的演讲《我有一个梦想》：

我有一个中国梦，人与自然能和谐相处的梦。

In spring, there will be a rapid cooling in Wuhan every year.
In the morning, a less active damselfly stayed on the dandelion.
In order not to startle it, I lay on the ground and quietly approached it.
I wanted to take a positive picture of the damselfly.
At this time, it raised a forefoot to wipe its eyes.
I recorded this moment with high-speed continuous photography.
When I got home, I picked out this one from several high-speed photos.
At that moment, the damselfly front feet staying in the air is like a gesture improvised by a speaker.
It is delivering the famous speech “I have a dream”:
I have a Chinese dream, that is man and nature can live in harmony.

我有一个梦想
I Have a Dream

雨后放晴，一只小蜗牛在寻觅它的栖身之所，

它爬上了苏铁的嫩叶。

苏铁嫩叶顶端卷曲的形态形成一种音乐旋律的节奏。

叶片顶端的螺旋形和蜗牛壳的螺旋形形成了一种韵律和对比。

It clears up after the rain.

A little snail is looking for its shelter, and it climbs tender leaves of the cycad (*Cycas revoluta*).

The curly shape at the top of tender leaves of the cycad forms a rhythm of music melody.

The spiral shape at the top of the leaf forms a rhythm and contrast with the spiral shape of the snail shell.

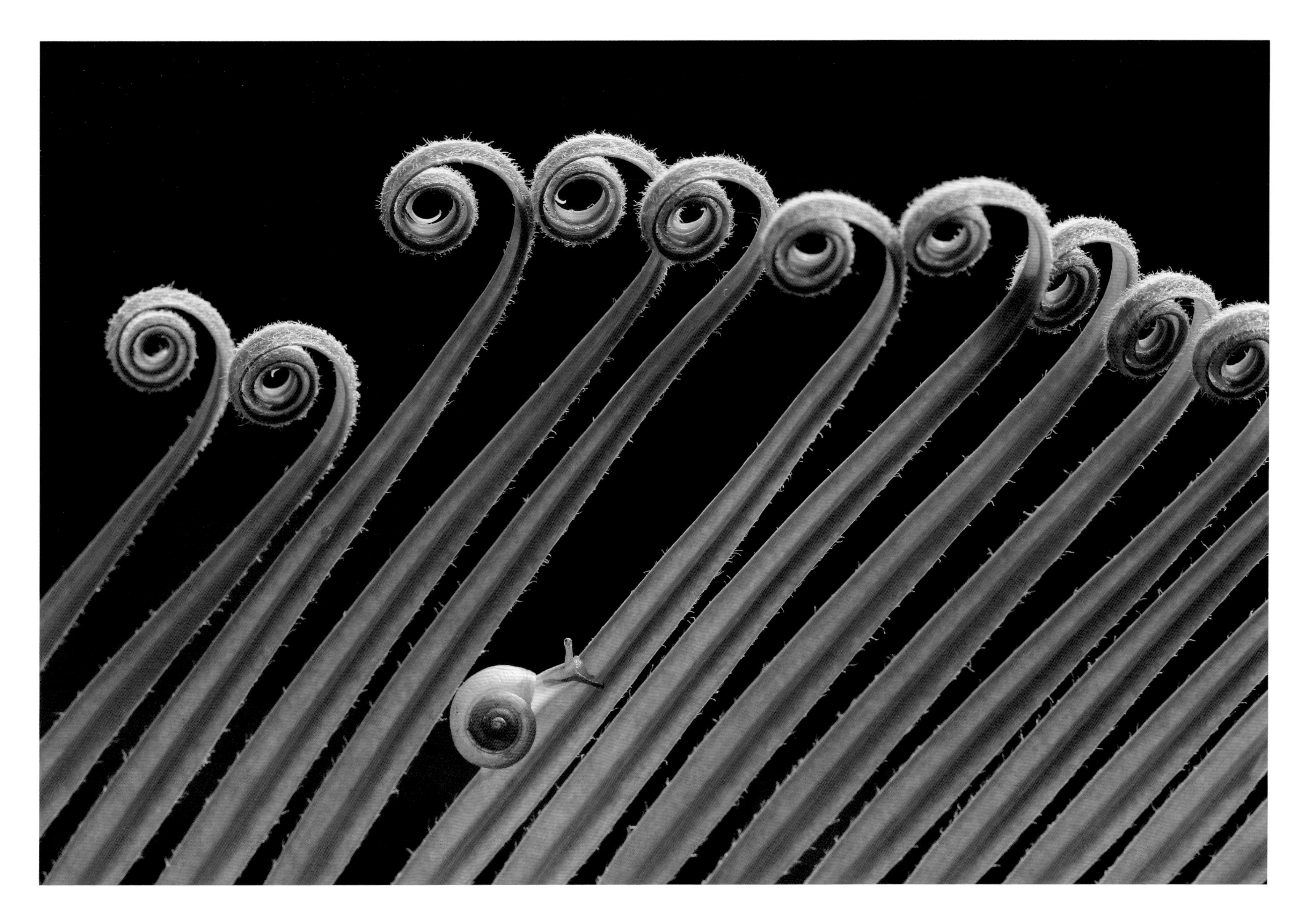

螺旋的韵律
Spiral Rhythm

在武汉市的夏季，我站在一个高的地方用长焦镜头拍摄东湖的荷叶。

水面的反光和镜像原理让画面变得简洁。

高高的荷叶挺立在小荷花的旁边。

荷叶在呵护下面的小荷花成长，它就像一把伞，为下面的小荷花遮风挡雨。

我想象中的小荷花就是一个小姑娘，她生活在妈妈的保护伞之下。

In the summer of Wuhan East Lake wetland,
I stood in a high place and photographed the lotus leaves with a telephoto lens in the East Lake.
The principle of reflection and mirror image on the water surface makes the picture concise.
The tall lotus leaves stand beside the small lotus.
The lotus leaf is taking care of the little lotus growing below.
It is like an umbrella to protect the little lotus from the wind and rain.
The little lotus in my imagination is a little girl who lives under her mother’s umbrella.

为你打伞
Umbrella for You

每年夏季，在湖北大别山溪水边，每隔三四米就会有一只透顶单脉色蟌。
雄性透顶单脉色蟌喜欢划地为王，任何飞临它领地的雄性同类都会被它驱赶，
但如果有雌性色蟌飞过它的领地，它会迅速起飞并开始热情地求偶。
雄性色蟌全身散发出强烈的金属光泽，在逆光下看却是黑色。
炎热的夏季，我穿着连体的防水衣裤跪在冰冷的溪流中，
把相机贴近水面，从下往上以天空为背景拍摄这只色蟌。
草叶简洁的线条和色蟌形成了线和点的呼应，就像一幅简约的中国画。
自然本身就是一幅简单的画作。

Every summer, single-veined pierogi,
who lives by the stream of Dabie Mountain in Hubei Province, has one every 3 or 4 meters.
The male likes to delimit the land as the king,
and any male who flies to its territory will be driven away by it,
but if a female color beetle flies over its territory, it will take off quickly and start courtship enthusiastically.
Male single-veined pierogi exudes a strong metallic luster, but it is black in the backlight.
In the hot summer, I knelt in the cold stream in one-piece waterproof clothes and trousers.
I put my camera close to the water and shot this single-veined pierogi with the sky as the background from bottom to top.
The simple lines of grass leaves and single-veined pierogi form the echo of lines and points,
which is like a simple Chinese painting.
Nature itself is a simple painting.

中国画
Chinese Painting

清晨，武汉市黄陂区的大别山边，
一只盲蛛倒挂在草叶上清洁整理自己的长腿。
这种不吐丝的蜘蛛，多出没于潮湿的泥土面和草丛。
此刻，一只小蜘蛛也爬上了草叶，
在盲蛛的长腿之下，它的体形和盲蛛形成了大与小的对比。

In the morning, on the edge of Dabie Mountain in Huangpi District,
Wuhan, a harvestman (Opiliones) hung upside down on grass leaves to clean and tidy up its long legs.
This kind of spider that doesn’t spin silk mostly haunts the wet soil surface and grass.
At this moment, a little spider also climbed up the grass leaves.
It was located under long legs of the harvestman.
Its size is in contrast to that of the harvestman.

长腿如丝
Long Legs like Silk

一只青蛙在露出水面的睡莲嫩叶的中间露出了头。

当我趴在地上拍摄时，我可以用一个平视的角度正面拍摄青蛙的眼睛。

通过水面的倒影我看到有趣的场景：

钻出水面的睡莲叶像是衬衫的衣领，小青蛙的头从衬衫的领口里钻出，

衬衫领口还有着一个领结，而这个领结是光线经过水面三次折射和反射后的一个倒影。

我想用最直接简单的构图方式，让观者回忆童年的轻松和诙谐，感受自然带来的快乐和真诚。

A frog shows its head in the middle of tender leaves of the water lily.
When I shot on the ground, I could shoot the frog's eyes from a head up angle.
Through the reflection of the water,
I saw an interesting scene: the water lily leaves drilling out of the water looked like the collar of a shirt.
The little frog's head came out of the collar of his shirt, that also had a bow tie.
This bow tie was a reflection of the light refracted and reflected three times through the water.
I want to use the most direct and simple composition to let the viewer recall the lightness and humor of childhood and feel the happiness and sincerity brought by nature.

青蛙王子的领结
Frog Prince’s Bow Tie

获奖作品

The Award-winning Works

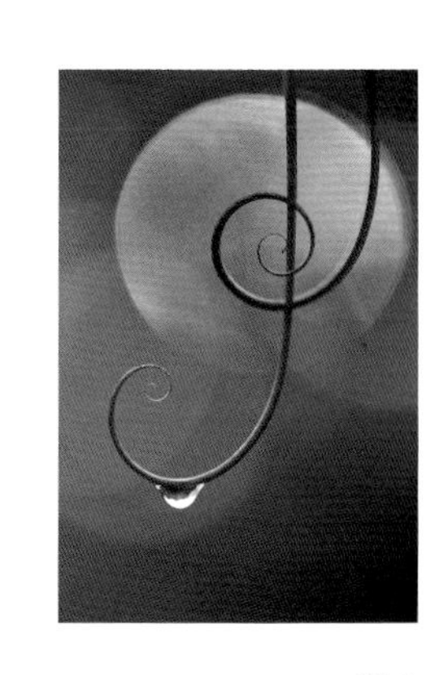

自然的和声

Natural Harmony

P17

2014年第50届国际野生生物摄影年赛植物和真菌组提名奖。
2014年第18届世界最佳自然摄影奖自然艺术组高度赞扬奖。
2015年第17届国际自然摄影亮点奖植物和真菌组高度赞扬奖。
2015年第25届国际山地与自然摄影年赛植物组高度赞扬奖。
2016年第16届中国国际摄影艺术展入选。

Nomination Award for plants and fungi group of the 50th International Wildlife Photographer of the Year (WPY) in 2014.
Highly Commended Award for nature art group of the 18th Nature's Best Photography International Awards (Windland Smith Rice) in 2014.
Highly Commended Award for plants and fungi group of the 17th International Nature Photography Award (Glanzlichter) in 2015.
Highly Commended Award for plants group of the 25th Memorial María Luisa in 2015.
Selected in the 16th China International Photographic Art Exhibition in 2016.

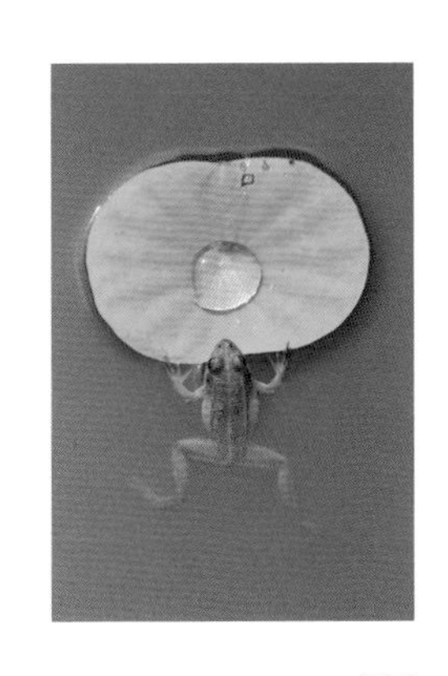

等待进餐
Waiting for Dinner

P25

2014年第7届国际园艺摄影年赛花园中的野生动物组高度赞扬奖。
2015年第1届亚洲最佳自然摄影奖微距世界组高度赞扬奖。
2021年第3届国际近摄特写摄影大赛动物组提名奖。

Highly Commended Award for wildlife in the garden group of the 7th International Garden Photographer of the Year (IGPOTY) in 2014.
Highly Commended Award for macro world group of the 1th Nature's Best Photography Asia Awards in 2015.
Nomination Award for the animal group of the 3rd Close-up Photographer of the Year in 2021.

眼中的一滴泪
A Tear in the Eye

P27

2020年第22届国际自然摄影亮点奖植物和真菌组最高荣誉奖。
2019年第2届西班牙卡迪斯国际自然摄影大赛植物和真菌组高度荣誉奖。

Highest Honor Award for plants and fungi group of the 22nd International Nature Photography Award (Glanzlichter) in 2020.
High Honor Award for plants and fungi group of the 2nd International Nature Photography Competition (Cádiz Photo Nature) in 2019.

风雨后相拥
Embracing Each Other after Rain

P29

2015年第1届亚洲最佳自然摄影奖微距世界组高度赞扬奖。

Highly Commended Award for macro world group of the 1th Nature's Best Photography Asia Awards in 2015.

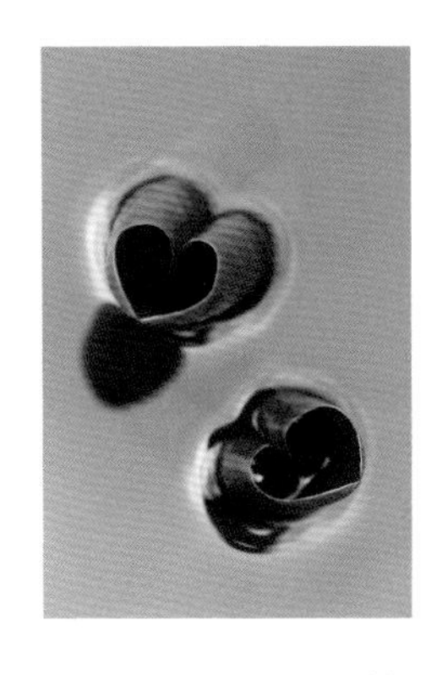

漂浮的心 Floating Heart

P33

2014年第7届国际园艺摄影年赛植物与水组赞扬奖。

Commended Award for plants & water group of the 7th International Garden Photographer of the Year (IGPOTY) in 2014.

毛毛虫的理想 Caterpillar's Ideal

P41

2013年第18届世界最佳自然摄影奖微距世界组高度赞扬奖。
2013年第6届国际园艺摄影年赛花园中的野生动物组提名奖。
2017年第2届荷兰国际自然摄影年赛其他动物组提名奖。
2017年第3届亚洲最佳自然摄影奖微距世界组高度赞扬奖。

Highly Commended Award for macro world group of the 18th Nature's Best Photography International Awards (Windland Smith Rice) in 2013.
Nomination Award for wildlife in the garden group of the 6th International Garden Photographer of the Year (IGPOTY) in 2013.
Nomination Award for other animals group of the 2nd Netherlands International Nature Photographer of The Year (NPOTY) in 2017.
Highly Commended Award for macro world group of the 3rd Nature's Best Photography Asia Awards (NBPA) in 2017.

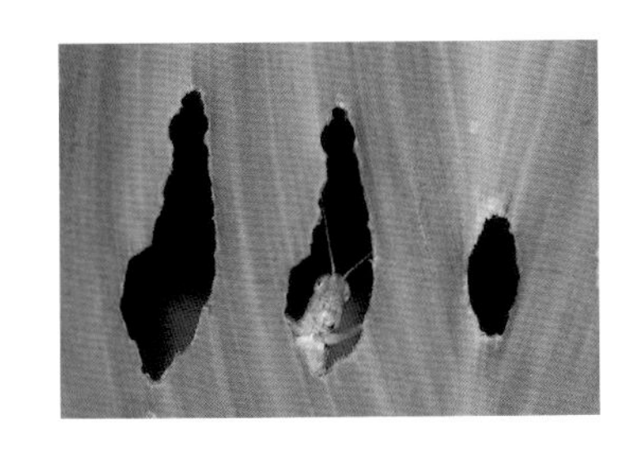

捉迷藏 Hide and Seek

P43

2012年第42届美国国家野生动物摄影大赛其他动物组高度赞扬奖。
2011年美国国家地理全球摄影大赛中国区自然组三等奖。
2014年第19届世界最佳自然摄影奖动物滑稽组高度赞扬奖。

Highly Commended Award for other animals group of the 42nd National Wildlife Photo Contest (NWPC) in 2012.
The Third Award for China Nature Group of the National Geographic Global Photography Competition in China in 2011.
Highly Commended Award for funny animals group of the 19th Nature's Best Photography International Awards (Windland Smith Rice) in 2014.

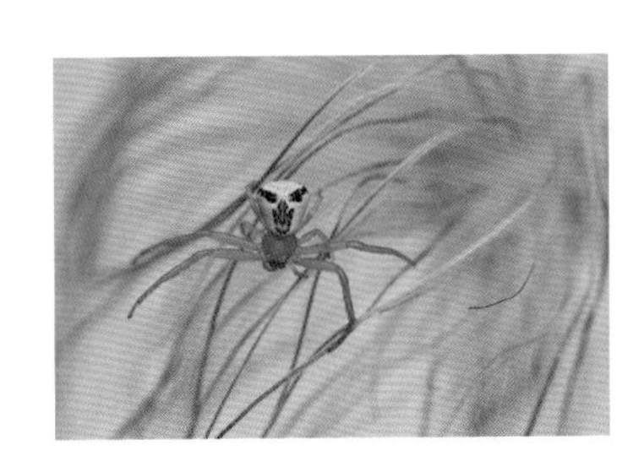

马戏小丑文身 Circus Clown Tattoo

P45

2014年BBC野生动物摄影杂志昆虫组月赛第一名。
2019年第6届意大利国际生物摄影师大赛其他动物组提名奖。

The First Award for the monthly competition of insect group of BBC Wildlife Photography Magazine in 2014.
Nomination Award for other animals group of the 6th Italian International Bio-photographer Contest in 2019.

在此等候 Right Here Waiting

P49

2022年第24届国际自然摄影亮点奖自然艺术组最高荣誉奖。

Highest Honor Award for nature art group of the 24th International Nature Photography Award (Glanzlichter) in 2022.

找个台阶下

Find a Step Down

P51

2016年第9届国际园艺摄影年赛花园中的野生动物组高度赞扬奖。
2016年美国最佳后院摄影奖冠军。

Highly Commended Award for wildlife in the garden group of the 9th International Garden Photographer of the Year (IGPOTY) in 2016.
Winner of the America Best Backyard Photography Award in 2016.

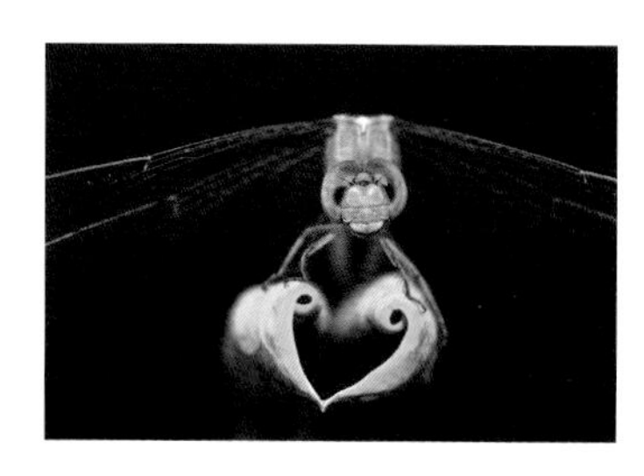

蜻蜓秀爱心

A Dragonfly Shows Love Heart

P53

2016年第1届国际自然摄影年赛其他动物组高度赞扬奖。
2017年第27届西班牙国际山地与自然摄影年赛微距组高度赞扬奖。

Highly Commended Award for other animals group of the 1st International Nature Photographer of The Year (NPOTY) in 2016.
Highly Commended Award for macro group of the 27th Memorial María Luisa in 2017.

蓝眼睛和绿眼睛

Blue and Green Eyes

P55

2020年第13届国际园艺摄影年赛微距艺术组提名奖。
2020年第2届国际近摄特写摄影大赛昆虫组提名奖。

Nomination Award for macroart group of the 13rd International Garden Photographer of the Year (IGPOTY) in 2020.
Nomination Award for insect group of the 2nd Close-up Photographer of the Year (CUPOTY) in 2020.

小四眼 P57

Little Four Eyes

2013年第6届国际园艺摄影年赛花园中的野生动物组高度赞扬奖。

Highly Commended Award for wildlife in the garden group of the 6th International Garden Photographer of the Year in 2013.

兄弟连 P59

Band of Brothers

2016年第6届国际自然影像奖其他动物组提名奖。
2017年第10届国际园艺摄影年赛花园中的野生动物组赞扬奖。

Nomination Award for other animals group of the 6th Nature Images Awards in 2016.
Highly Commended Award for wildlife in the garden group of the 10th International Garden Photographer of the Year in 2017.

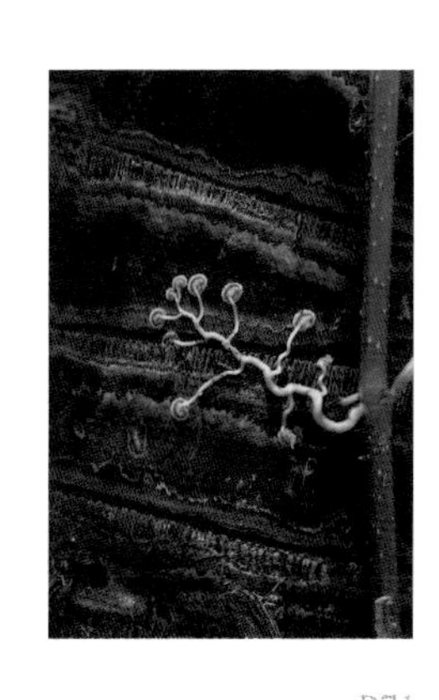

坚持 P71

Holding On

2018年第12届国际自然摄影竞赛奖植物和真菌组最高荣誉奖。
2018年第20届国际自然摄影亮点奖植物和真菌组高度赞扬奖。
2018年第11届国际园艺摄影年赛美丽植物组荣誉奖。

Highest Honor Award for plants and fungi group of the 12nd International Nature Photography Award (Asferico) in 2018.
Highly Commended Award for plants and fungi Group of the 20th International Nature Photography Award (Glanzlichter) in 2018.
Award of honor for beauty of plants group of the 11st International Garden Photographer of the Year in 2018.

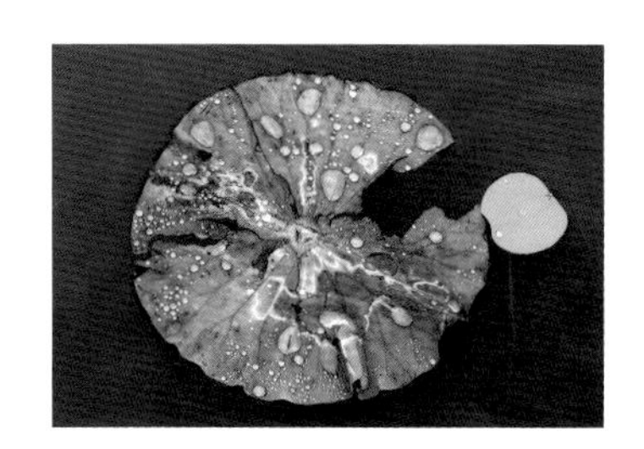

生生不息
The Circle of Life

P73

第7届国际园艺摄影年赛植物与水组提名奖。

Nomination Award for plants & water group of the 7th International Garden Photographer of the Year (IGPOTY) in 2014.

水中四季
Four Seasons in Water

P77

2018年第11届国际园艺摄影年赛微距艺术组荣誉奖。

Award of honor for macro art of the 11st International Garden Photographer of the Year (IGPOTY) in 2018.

倒影的反思
Reflection on Reflection

P79

2013年第43届国家野生动物摄影大赛景观与植物组高度赞扬奖。
2016年第1届国际自然摄影年赛植物和真菌组提名奖。
2019年第6届国际生物摄影师大赛淡水植物组高度赞扬奖。

Highly Commended Award for landscape & plants group of the 43rd National Wildlife Photo Contest (NWPC) in 2013.
Nomination Award for plants and fungi group of the 1st Nature Photographer of The Year (NPOTY) in 2016.
Highly Commended Award for fresh-water plants group of the 6th BioPhotoContest in 2019.

经历曲折后的重逢 Reunion after Twists and Turns

P81

2014年第7届国际园艺摄影年赛美丽植物组高度赞扬奖。

Highly Commended Award for beauty of plants group of the 7th International Garden Photographer of the Year (IGPOTY) in 2014.

女士 Lady

P87

2020年第13届国际园艺摄影年赛黑白组高度赞扬奖。

Highly Commended Award for black & white group of the 13rd International Garden Photographer of the Year (IGPOTY) in 2020.

水中的依偎 Snuggling in the Water

P91

2017年第10届国际园艺摄影年赛微距艺术组提名奖。

Nomination Award for macro art group of the 10th International Garden Photographer of the Year (IGPOTY) in 2017.

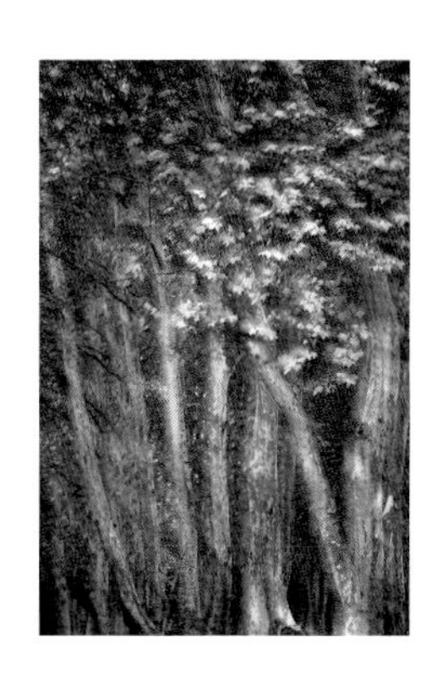

秋日印象
Autumn Impression

P107

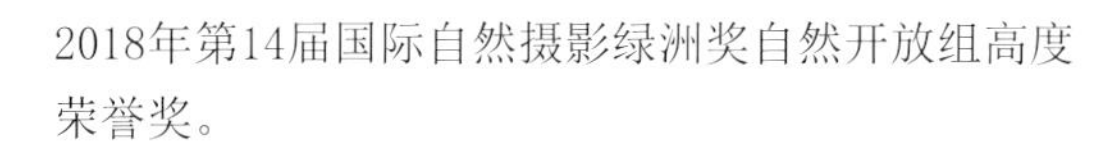

2018年第14届国际自然摄影绿洲奖自然开放组高度荣誉奖。

Highly Commended Award for natural open group of the 14th International Nature Photography Award (OASIS) in 2018.

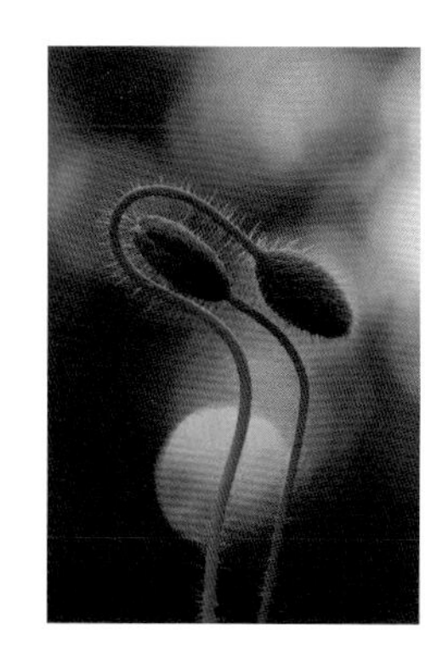

阳光下的拥抱
Embrace in the Sun

P111

2017年第10届国际园艺摄影年赛美丽植物组提名奖。

Nomination Award for beauty of plants group of the 10th International Garden Photographer of the Year (IGPOTY) in 2017.

贴面舞
Cheek-to-cheek Dancing

P113

2016年第18届国际自然摄影亮点奖植物和真菌组高度赞扬奖。

Highly Commended Award for plants and fungi group of the 18th International Nature Photography Award (Glanzlichter) in 2016.

青苹果乐园
Green Apple Paradise

P147

2017年第3届亚洲最佳自然摄影奖微距世界组高度赞扬奖。

Highly Commended Award for macro world group of the 3rd Nature's Best Photography Asia Awards (NBPA) in 2017.

世界之巅
Top of the World

P149

2016年第2届亚洲最佳自然摄影奖微距世界组高度赞扬奖。

Highly Commended Award for macro world group of the 2nd Nature's Best Photography Asia Awards (NBPA) in 2016.

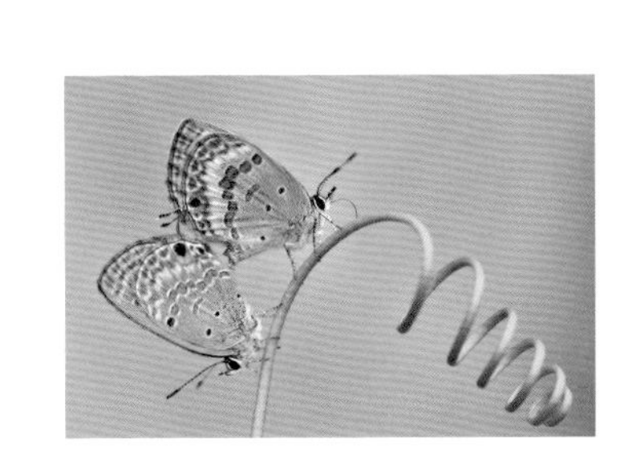

螺旋的爱
Spiral Love

P157

2014年美国最佳后院摄影奖冠军。
2015年第25届西班牙国际山地与自然摄影年赛微距组高度赞扬奖。

The First Award for the America Best Backyard Photography Award in 2014.
Highly Commended Award for the macro group of the 25th Memorial María Luisa in 2015.

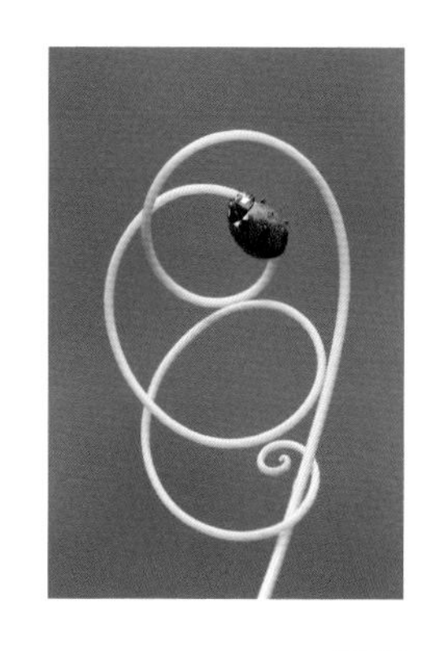

生活就像过山车 P171
Life Is like a Roller Coaster

2015年第1届亚洲最佳自然摄影奖微距世界组冠军。
2015年第8届国际园艺摄影年赛花园中的野生动物组高度赞扬奖。
2020年第22届国际自然摄影亮点奖其他动物组最高荣誉奖。

The First Award for macro world group of the 1st Nature's Best Photography Asia Awards (NBPA) in 2015.
Highly Commended Award for wildlife in the garden group of the 8th International Garden Photographer of the Year (IGPOTY) in 2015.
Highest Honor Award for other animals Group of the 22nd International Nature Photography Award (Glanzlichter) in 2020.

射灯下的独舞 P173
Solo Dance under Spotlight

2018年第11届国际园艺摄影年赛花园中的野生动物组高度赞扬奖。

Highly Commended Award for wildlife in the garden group of the 11st International Garden Photographer of the Year (IGPOTY) in 2018.

打伞的豆娘 P175
Damselfly with Umbrella

2019年第12届国际园艺摄影年赛黑白组第三名。
2015年美国最佳后院摄影奖冠军。

The Third Award for black & white group of the 12nd International Garden Photographer of the Year (IGPOTY) in 2019.
The First Award for the America Best Backyard Photography Award in 2015.

思想者 Thinker

P139

2014年世界最佳自然摄影奖自然艺术组高度赞扬奖。
2015年第1届亚洲最佳自然摄影奖微距世界组高度赞扬奖。

Highly Commended Award for nature art group of the Nature's Best Photography International Awards (Windland Smith Rice) in 2014.
Highly Commended Award for macro world group of the 1st Nature's Best Photography Asia Awards (NBPA) in 2015.

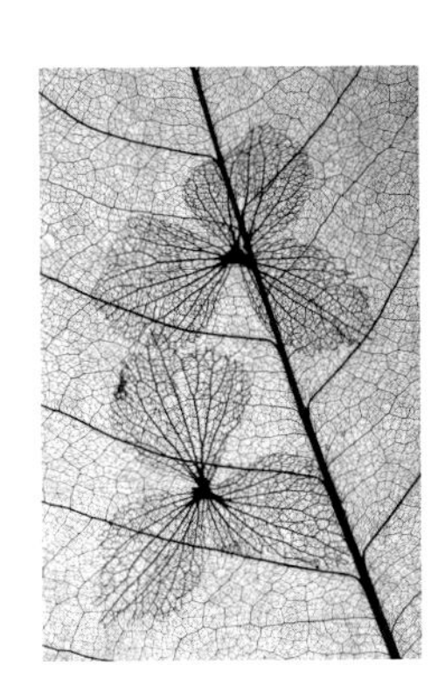

蕾丝 Lace

P179

2018年第11届国际园艺摄影年赛微距艺术组高度荣誉奖。
2018年第20届国际自然摄影亮点奖自然艺术组最高荣誉奖。

Highly Commended Award for macro art group of the 11st International Garden Photographer of the Year in 2018.
Highest Honor Award for nature art group of the 20th International Nature Photography Award (Glanzlichter) in 2018.

我有一个梦想 I Have a Dream

P183

2015年美国国家地理全球摄影大赛中国组自然组冠军。

The First Award for China nature group of the America National Geographic in 2015.

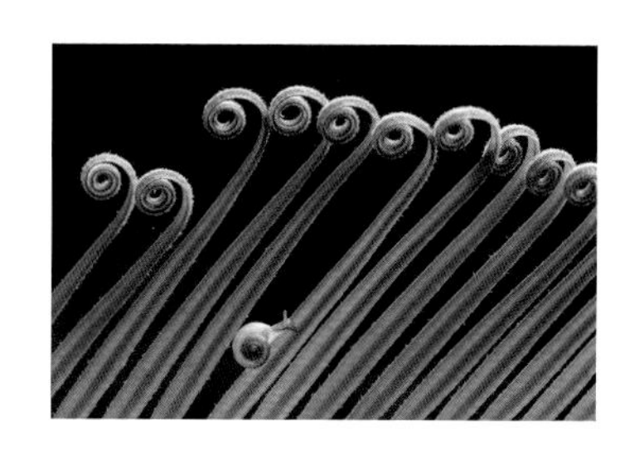

螺旋的韵律
Spiral Rhythm

P185

2015年第11届国际自然摄影绿洲奖其他动物组高度荣誉奖。
2015年美国最佳后院摄影奖冠军。

Highly Commended Award for other animals of the 11st International Nature Photography Award (OASIS) in 2015.
The First Award for the America Best Backyard Photography Award in 2015.

为你打伞
Umbrella for You

P187

2017年第11届国际自然摄影竞赛奖植物和真菌组冠军。
2017年第10届国际园艺摄影年赛黑白组季军。
2017年第21届国际自然与野生动物摄影节植物和真菌组高度荣誉奖。

The First Award for plants and fungi group of the 11st of International Photography Competition Award (Asferico) in 2017.
The Third Award for balck & white group of the 10th International Garden Photographer of the Year (IGPOTY) in 2017.
Highly Commended Award for plants and fungi group of the 21st Montier Festival Photo in 2017.

中国画
Chinese Painting

P189

2020年第1届TTL自然摄影年赛微距组冠军.
2014年第7届国际园艺摄影年赛黑白组提名奖。

The First Award for macro group of the 1st TTL Nature Photography of the Year (POTY) in 2020.
Nomination Award for black & white group of the 7th International Garden Photographer of the Year (IGPOTY) in 2014.

P193

青蛙王子的领结
Frog Prince's Bow Tie

2016年第18届国际自然摄影亮点奖其他动物组冠军。
2016年美国最佳后院摄影奖冠军。
2016年第2届亚洲最佳自然摄影奖微距世界组冠军。
2017年第21届国际自然与野生动物摄影节其他动物组冠军。
2017年第10届国际园艺摄影年赛花园中的野生动物组提名奖。
2017年第13届国际自然摄影绿洲奖其他动物组荣誉奖。
2017年第11届国际野生动物艺术与摄影大赛动物肖像组提名奖。
2018年第28届国际山地与自然摄影年赛微距组高度赞扬奖。

The First Award for other animals group of the 18th International Nature Photography Award (Glanzlichter) in 2016.
The First Award for the America Best Backyard Photography Award in 2016.
The First Award for macro world group of the 2nd Nature's Best Photography Asia Awards (NBPA) in 2016.
The First Award for other animals group of the 21st Montier Festival Photo in 2017.
Nomination Award for wildlife in the garden group of the 10th International Garden Photographer of the Year (IGPOTY) in 2017.
Highly Commended Award for other animals of the 13rd International Nature Photography Award (OASIS) in 2017.
Nomination Award for animals portraits group of the 11st International Wildlife Art and Photography Competition (Golden Turtle) in 2017.
Highly Commended Award for macro group of the 28th Memorial María Luisa in 2018.

艺术成就
Artistic Achievements

1. 获得三次（2014、2019、2021年）年度野生动物摄影师（WPY）奖项。
2. 国际花园摄影年赛（IGPOTY）历史上获奖的第一位中国摄影师，第一位获得组别冠军的亚洲摄影师，也是国际花园摄影年赛历史上获奖最多的摄影师。
3. 英国TTL国际自然摄影年赛（POTY）历史上第一个冠军，也是历史上获奖的第一位亚洲的摄影师。
4. 国际近摄特写摄影大赛（CUPOTY）历史上获奖的第一位中国摄影师。
5. 美国国家野生动物摄影大赛（NWPC）历史上获奖的第一位中国摄影师。
6. 世界最佳自然摄影奖（Windland Smith Rice）历史上获奖的第一位中国摄影师。
7. 美国最佳后院摄影奖（Best Backyards）历史上获奖的第一位中国摄影师，单届比赛四张作品并列冠军，为美国最佳后院摄影奖历史纪录。
8. 亚洲最佳自然摄影奖（NBPA）历史上第一个冠军，也是获奖的第一位中国摄影师，该赛事历史上获奖最多的摄影师。

9. 美国园艺摄影大赛(Horticulture Garden)历史上获奖的第一位中国摄影师。

10. 德国国际自然摄影亮点奖（Glanzlichter）历史上获奖的第一位中国摄影师，也是第一个获得组别冠军的亚洲摄影师，该赛事历史上获奖最多的摄影师。

11. 法国国际自然与野生动物摄影节(Montier Festival Photo)历史上获奖的第一位中国摄影师,也是第一个获得组别冠军的亚洲摄影师。

12. 法国国际自然影像奖(Nature Images Awards)历史上获奖的第一位中国摄影师。

13. 国际自然摄影竞赛奖(Asferico)历史上获奖的第一位中国摄影师,也是第一个获得组别冠军的亚洲摄影师。

14. 意大利国际自然摄影绿洲奖(OASIS)历史上获奖的第一位中国摄影师。

15. 意大利国际生物摄影师大赛(BioPhotoContest)历史上获奖的第一位中国摄影师。

16. 西班牙国际山地与自然摄影年赛(Memorial María Luisa)历史上获奖的第一位中国摄影师。

17. 西班牙国际自然摄影大赛（Montphoto）历史上获奖的第一位中国摄影师。

18. 西班牙国际自然摄影大赛(Cádiz Photo Nature)历史上获奖的第一位中国摄影师,第一个获得组别冠军的亚洲摄影师。

19. 荷兰国际自然摄影年赛(NPOTY)历史上获奖的第一位中国摄影师。

20. 俄罗斯国际野生动物艺术与摄影大赛(Golden Turtle)历史上获奖的第一位中国摄影师。

1. Won the Wildlife Photogarapher of the Year (WPY) award three times (2014, 2019, 2021).

2. The first Chinese photographer to win the prize in the history of the International Garden Photography Competition (IGPOTY), the first Asian photographer to win the group champion, and the photographer with the most awards in the history of IGPOTY.

3. The first champion in the history of the British TTL International Nature Photography Competition (POTY), and the first Chinese and Asian photographer to win the award.

4. The first Chinese photographer to win an award in the history of the International Close-up Photographer of the Year (CUPOTY).

5. The first Chinese photographer to win an award in the history of the National Wildlife Photo Contest (NWPC).

6. The first Chinese photographer in the history of the World Best Nature Photography Award (Windland Smith Rice).

7. The first Chinese photographer to win the Best Backyard Photography Award in the history of the United States, and the four works tied for the championship in a single competition are the historical record of the Best Backyard Photography Award in the United States (Best Backyards).

8. The first champion in the history of the Best Nature Photography Awards in Asia (NBPA), and the first Chinese photographer to win the award, the photographer with the most awards in the history of the event.

9. The first Chinese photographer to win an award in the history of the Horticulture Garden competition.

10. The first Chinese photographer to win the Glanzlichter in the history of Germany (Glanzlichter), and the first Asian photographer to win the category championship, the photographer with the most awards in the history of the competition.

11. The first Chinese photographer to win the prize in the history of the Montier Festival photo in France (Montier Festival photo) , and also the first Asian photographer to win the group championship.

12. The first Chinese photographer to win the French nature images awards (Nature images awards).

13. The first Chinese photographer in the history of the international natural Photography Competition Award (Asferico) and the first Asian photographer to win the category champion.

14. The first Chinese photographer to win the oasis Award for natural photography (OASIS) in the history of Italy.

15. The first Chinese photographer in the history of the Italian international BioPhotoContest (BioPhoto).

16. The first Chinese photographer in the history of the memorial María Luisa (MML), Spain.

17. The first Chinese photographer to win the prize in the history of the Spanish international natural photography competition (Montphoto).

18. The first Chinese photographer and the first Asian photographer to win the category championship in the history of the Spanish International Nature Photography Competition (Cádiz Photo Nature).

19. The first Chinese photographer to win the prize in the history of the Netherlands international natural photography annual competition (NPOTY).

20.The first Chinese photographer in the history of the Russian International Wildlife Art and photography competition (Golden Turtle).

后记
Epilogue

生命中始终有一种距离，能够退后让你静心思考你想做什么。这时我发现，如果把自己的兴趣爱好和工作结合起来则是最幸福的事。生在中国的我有一个中国梦！我的梦想是成为一名自然生态摄影师。

我在2014年，首次获得年度野生动物摄影师（WPY）奖项后觉得，是时候为自己所爱的自然生态摄影奉献自己的全部。2015年，我辞职后全身心投入我热爱的事业，有了更多的时间表现我对生命的理解。我用自然摄影这种方式向世界展示中国的自然之美，让世界了解中国人对生命的尊重，知道中国人和世界各民族的人一样对世界充满了爱的情感！

中国现在把湿地保护摆上更加突出的位置，与经济社会发展各项任务统筹考虑，落实好湿地保护责任，建立湿地保护长效机制，强化宣传教育，提高全民湿地保护意识。中国在不断扩大湿地面积，增强湿地生态系统稳定性，进一步改善生态和民生。

这本《微距生灵》是大自然给我的启迪。不仅是我拍摄湿地的心灵之旅，也是我一直以来的梦想。看到这些作品，我依然能回忆起当时的发现和拍摄时的情景，就像我在梦里依然在摄影。在梦中，意识回归到最原始的状态。在意识转换的一刹那，“有我”与“无我”、“今日之我”和“昨日之我”彻底决裂，从而产生创意。思维模式和视觉习惯被打破，每一刻都会是崭新的。每一次的创作都是在尝试一个新鲜的感觉。作为一个自然的欣赏者，我在摄影中感悟自然之道、生命的哲学，慢慢地融入自然，包括：大树、昆虫、野兽、飞鸟、游鱼……多种生命的体验。生与死、追逐与逃离、孤独与陪伴、爱与被爱、顺服与抗争……对其他生命的痛苦和欢乐感同身受。

在大自然中，我经历了独特与不平凡，又回归了纯粹与质朴。大自然的每个方寸之地都隐藏着星辰大海般的秘密，一片落叶可以述说一棵树的过去，树豆荚里的种子可以展示这片土地的未来……自然摄影是创造属于你的情感世界的艺术，在影像中寻找人性最细腻的情感。这时候，我们拥有完全的自由，恢复了生命的完整。自然摄影艺术将你的激情和感受广泛地传递，在对生命的了解达到新的高度后，你的生命力会重新凝聚而产生新的动力。

我知道我参与自然发现的伟大探索永远不会简单地结束，它将继续呈现出更加美妙的形式。诗歌与音乐都可以给我一个意象，让我用摄影向自然做一个祈祷。

袁明辉

2022年9月

There is always a distance in life, which can make you step back and meditate. What do you want to do? At this time, I find that it is the happiest thing to combine my hobbies with my work. Born in China, I have a Chinese dream! My dream is to become a natural ecology photographer.

In 2014, after I won the Wildlife Photographer of the Year Award for the first time, I felt it was time to devote myself to my beloved natural ecology photography. In 2015, I resigned and devoted myself to the cause I loves. So, I had more time to express my understanding of life. I use natural photography to show the world the natural beauty of China. Let the world know the Chinese people's respect for life. Chinese people and people of all nationalities in the world are also full of love for the world!

China has now placed wetland protection in a more prominent position, taking into account all tasks related to economic and social development and implementing the responsibility of wetland protection. China has established a long-term mechanism for wetland protection, strengthened publicity and education, and improved the wetland protection awareness of the whole people. China is constantly expanding the wetland area, enhancing the stability of the wetland ecosystem, and further improving the ecology and people's livelihood.

This *Close-up Views of Creatures in Wedlands* is the enlightenment of nature to me. It is not only my spiritual journey to shoot wetlands, but also my dream all the time. Seeing these works, I can still recall the discovery and shooting scenes at that time. It's like I'm still photographing in my dream. In dreams, consciousness returns to its most primitive state. At the moment of consciousness transformation, “self ” and “no self ”, “today’s self ” and “yesterday’s self ” completely break away, thus creating creativity. Thinking patterns and visual habits are broken, and every moment will be new. Every time I create, I try a new feeling. As a natural admirer, I feel the way of nature and the philosophy of life in photography, and slowly integrate into nature, including: trees, insects, beasts, birds, fish, ··· A variety of life experiences. Life and death, chasing and escaping, loneliness and company, love and being loved, obedience and resistance ··· I feel the pain and joy of other lives.

In nature, I experienced uniqueness and extraordinary, and returned to purity and simplicity. Every inch of nature hides the secrets of the stars and the sea. A fallen leaf can tell the past of a tree, and the seeds in the tree pods can show the future of this land··· Natural photography is the art of creating your own emotional world, looking for the most delicate emotion of human nature in the image. At this time, we have complete freedom and restore the integrity of life. The art of natural photography transmits your passion and feelings widely. After your understanding of life reaches a new level, your vitality will condense again and generate new power.

I know that my great exploration of natural discovery will never end simply, and it will continue to take on a more wonderful form. Poetry and music can give me an image. Let me pray to nature with photography.

Yuan Minghui

September, 2022